ADVANCES IN PIEZOELECTRIC TRANSDUCERS

Edited by **Farzad Ebrahimi**

Advances in Piezoelectric Transducers
Edited by Farzad Ebrahimi

Published by InTech
Janeza Trdine 9, 51000 Rijeka, Croatia

Publishing Process Manager Ivana Zec

Technical Editor Teodora Smiljanic

Cover Designer InTech Design Team

Additional hard copies can be obtained from orders@intechopen.com

Advances in Piezoelectric Transducers, Edited by Farzad Ebrahimi
p. cm.
ISBN 978-953-307-931-8

Meet the editor

Dr. Farzad Ebrahimi was born in Qazvin, Iran, in 1979. He graduated in Mechanical engineering, from the University of Tehran, Iran, in 2002. He received his Msc and PhD in Mechanical engineering, with a specialization in applied design from the University of Tehran, Iran, in 2009. Since 2002, he has been working at the "Smart Materials and Structures Lab" Research Center of the Faculty of Mechanical Engineering at the University of Tehran, where he is a researcher of smart functionally graded materials and structures. He is involved in several international journals as editor and reviewer. He serves on the editorial board of "Far East Journal of Mechanical Engineering and Physics". He is the author of books "Smart Functionally Graded Plates" and "Progress in Analysis of Functionally Graded Structures", Nova Science Publishers, NY. He also served as the editor of the book "Advances in Vibration Analysis Research", InTech – Open Access Publisher, Croatia. His research interests focus on areas of smart materials and structures, plate and shell theory, vibration analysis of continuous systems, composite materials and structures, functionally graded materials and structures, finite element analysis and fracture mechanics and he has published several researches in these fields. His research in these areas has been presented at international conferences and appeared in academic journals such as "Journal of Mechanical Science and Technology", "Smart Materials and Structures", "European Journal of Mechanics" and "Archive of Applied Mechanics and Thin Walled Structures". Farzad has also strong collaboration with Iranian industries on gas and oil projects, as well as serves as ad-hoc referee in several top academic journals.

Contents

Preface

A transducer is a device that converts energy from one form to another. Presently, piezoelectric material is commonly used as a basic component of transducers. Piezoelectric devices are a very reliable and inexpensive means of converting electrical energy into physical motion and exhibit a high tolerance to environmental factors such as electromagnetic fields and humidity. The piezoelectric transducer converts electric signals into mechanical vibrations or vice versa by utilizing the morphological change of a crystal which occurs on voltage application, or conversely by monitoring the voltage generated by a pressure applied on a crystal.

Piezoelectric transducers have many applications. In particular, piezoelectric diaphragms have been employed as pressure sensors, in speakers for audio equipment, fluid ejection, fluid pumping and printing applications. Piezoelectric ceramic transducers have also found many important applications in adaptive structures for vibration control and acoustic noise suppression in modern space, civilian and military systems, such as launch vehicles, space platforms, aircraft, submarines and helicopters.

This book reports on the state of the art research and development findings on this very broad matter through original and innovative research studies exhibiting various investigation directions.

In particular, it introduces recent research results on important issues at the field of piezoelectric transducers modeling and applications.

The book has been organized in two parts. The first part deals with the modeling approaches of the piezoelectric transducers e.g. modeling of the horn-type piezoelectric ultrasonic transducers in chapter 1 and the modeling and investigation of one-dimensional flexural vibrating mechatronic systems with piezoelectric transducers in chapter 2. Also, the distributed-parameter modeling of energy harvesting structures with discontinuities and the correcting magnitude and phase distortion in wideband piezoelectric transducer systems is introduced in chapter 3 and 4. In the second part, the applications of piezoelectric transducers in structural health monitoring and the data acquisition and virtual instrumentation for electromechanical impedance technique are introduced in chapters 5 and 6, respectively.

The advantage of the book "**Advances in Piezoelectric Transducers**" is its open access, fully searchable by anyone anywhere, and in this way it provides the forum for dissemination and exchange of the latest scientific information on theoretical as well as applied areas of knowledge in the field of piezoelectric transducers.

The present book is a result of contributions of experts from international scientific community working in different aspects of piezoelectric transducers. The introductions, data, and references in this book will help the readers know more about this topic and help them explore this exciting and fast-evolving field.

The text is addressed not only to researchers, but also to professional engineers, students and other experts in a variety of disciplines, both academic and industrial seeking to gain a better understanding of what has been done in the field recently, and what kind of open problems are in this area.

I hope that readers will find the book useful and inspiring by examining the recent developments in piezoelectric transducers.

Dr. Farzad Ebrahimi
Mechanical Engineering Department,
University of Tehran – Tehran,
I. R. Iran

Part 1

Modeling of Piezoelectric Transducers

1

Horn-Type Piezoelectric Ultrasonic Transducer: Modelling and Applications

Tao Li[1*], Jan Ma[1] and Adrian F. Low[2*]
[1]*School of Materials Science and Engineering, Nanyang Technological University*
[2]*National University Heart Centre; National University Health System, Singapore*

1. Introduction

The piezoelectric transducers can be categorized into sonic, ultrasonic and megasonic transducers based on the operating frequency. In each category, the design and function of the transducers vary significantly. The sonic transducers work at an audible frequency range, typically less than 20 kHz. In this frequency range, the transducers could be designed in the bending mode. Examples of the sonic transducers are the bimorph cantilever and buzzer (siren) unimorph transducer (APC International, Ltd., 2002; Yeo et al., 2008). The buzzer transducer is made of a piezoelectric disk attached to a metallic substrate. It is able to produce a sound level more than 100 dB at resonance in the frequency range of kHz (Bai et al., 2007; Miclea et al., 2008). At off-resonance conditions, the buzzer transducer can also be used as an actuator for piezoelectric diaphragm pump (Accoto et al., 2000; Woias, 2005). The ultrasonic transducers usually work at a frequency range from ~20 kHz to ~200 kHz (Abramov, 1998; Atar et al., 1999; Chen & Wu, 2002; Chu et al., 2002; Hongoh et al., 2004; Lin, 1995, 2004a; Mattiat, 1971; Medis & Henderson, 2005; Prokic, 2004; Radmanovic & Mancic, 2004; Sherrit et al., 1999a, 1999b; Tsuda et al., 1983; Wiksell et al., 2000). Transducers in this range could also be designed in the bending mode. Some examples are the tube transducer for cylindrical ultrasonic motor and disk/ring transducer for travelling wave motor (Li et al., 2007a; Uchino, 2003). But in this frequency range, a more common transducer design is the longitudinal mode. Langevin (sandwich, converter, Tonpiltz, etc.) transducer represents a typical structure (Abramov, 1998; Atar et al., 1999; Chen & Wu, 2002; Chu et al., 2002; Hongoh et al., 2004; Lin, 1995, 2004a, 2004b; Mattiat, 1971; Medis & Henderson, 2005; Prokic, 2004; Radmanovic & Mancic, 2004; Sherrit et al., 1999a, 1999b; Tsuda et al., 1983; Wiksell et al., 2000). It is sometimes connected with a horn (wave guide, booster, sonotrode, etc.) to focus or transmit energy. This type of transducer provides a wide range of applications, including welding, machining, sonochemistry, cleaning, underwater communication, ultrasonic surgery, etc. The megasonic transducers work in the frequency range of MHz. The most widely used vibration mode is the thickness mode for this range. One of the typical applications of the transducers in this category is the megasonic cleaning (Kapila et al., 2006). It provides advantages such as gentle and controllable cavitation, which will incur less damage on the cleaned parts compared with its traditional ultrasonic counterpart. Hence it is more suitable for precision cleaning.

---

* Corresponding Author

The capability of the transducer is closely associated with the frequency, vibration mode and wavelength. As a result, different transducers will achieve different performances in a wide range of applications. In comparison, the ultrasonic transducer with longitudinal vibration mode is the only one that is able to achieve both large vibration amplitude and high power (density) simultaneously. The vibration velocity of the transducer is limited by the mechanical, thermal and electrical properties of the materials. Currently, the PZT material is only able to achieve a vibration velocity less than 1 m/s as compared to 20 m/s for the titanium material in the same ultrasonic frequency range (Li et al., 2007b; Mason & Wehr, 1970; Muhlen, 1990). Thus, by using the titanium horn, the transducer can achieve larger vibration amplitude. In addition, the transducer at the axial direction has a relative long wave length compared to the thickness mode. A large volume of the PZT material could be used to increase the power capacity of the transducer (Mason, 1964). Furthermore, the power can be focused to a small area by the horn, which further improves the power density of the transducer. Due to the superiorities of high amplitude and high power (density), the horn type transducer has been found to have many unique applications.

The shape of the horn-type transducer varies significantly depending on the applications. Conventionally, the modelling of the transducer can be based on electrical circuit theory and FEA theory. Typically, the electrical circuit theory is used to design the transducer while FEA theory is used to analyze the performance of the transducer. It is convenient to use the electrical theory to design the dimension. However, for analysis of parameters, such as mode shape, stress distribution, power consumption, etc, the FEA theory is more advantageous. In the present work, we proposed an improved electrical circuit theory, called Finite Electrical Circuit Element Modelling (FECEM), which combines functions of both design and analysis. In this method, the transducer can be represented using a network, which are the connections of electrical circuit elements. The great advantage of the proposed method is the high efficiency in the development of piezoelectric transducers. We also proposed a method to estimate the mechanical quality factor Q_m which is important to estimate parameters like velocity and power consumption. Conventionally, Q_m parameter was either assumed or measured. In the present work, the FECEM was also compared with ANSYS FEA 2D analysis. Consistent results have been found.

The ANSYS 2D fluid-structure acoustic modelling was also introduced in the present work. In this modelling, the system response can be simulated when the transducer is loaded in the acoustic field. This model is useful to analyze the loading effect on the transducer. It is also essential to study the distributions and magnitude of sound pressure under different boundary conditions.

Lastly, in this chapter two applications of the horn-type transducer are introduced. They are thrombolysis transducer and acoustic pump. Both can be designed and analyzed using the above mentioned FECEM and ANSYS 2D theories. The thrombolysis transducer is a horn-type transducer connected with a long and thin transmission wire. It can be used to emulsify the thrombus in the blood vessel. The transducer was modelled and characterized in the present work. Efficient and effective emulsification results have been observed which proves the design and analysis were successful. The acoustic pump is another application of the horn-type transducer. The pump possesses a simple structure with just a transducer and a casing. Due to the vibration of the transducer, an acoustic field will be established inside the casing. As a result, the liquid will be forced to flow due to the nonlinear acoustic effects, such as streaming and cavitation. The unique features of the pump are high pressure head, continuous flow, no body oscillation and insensitive to the tubing length. The potential applications of the pump are sonochemistry and microfluidics.

2. General structure of the horn-type piezoelectric transducer

Fig. 1 shows the typical structure of the horn-type piezoelectric transducer. Generally, the transducer comprises four parts, i.e., the back mass, the piezoelectric stack, the front mass and the horn. The piezoelectric stack is clamped between the front and back mass. The horn is coupled to the front mass through the method such as bolt. In this configuration, the function of the piezo stack is to generate vibration, which is then amplified by the horn. The horn together with the front and back mass also has function of impedance match to maximize the energy transfer from the piezoelectric stack to the loadings at the tip of the transducer (Chen et al., 2004). The horn can have various designs based on the applications, such as conical profiled, step profiled, exponential profiled, etc (Hongoh et al., 2004; Horita, 1967; Khmelev et al., 2005; Minchenko, 1969; Muhlen, 1990). To maximize the vibration amplitude, the horn can be designed with multiple sections. Each section has the same or different profiles. The total length of the transducer is usually multiples of the half wavelength.

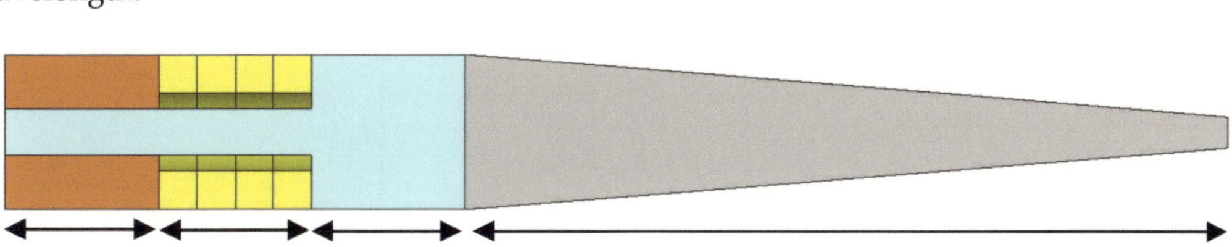

Fig. 1. General structure of a horn-type piezoelectric transducer

3. FECEM theory

Equivalent circuit network theory is a well-accepted method in piezoelectric material study and transducer design (Li et al., 2008, 2009, 2010; Mason,1964; Prokic, 2004; Radmanovic & Mancic, 2004). In this method, the transducer is divided into different sections. Each section will be represented using a two- or three-port electrical network. The networks are then connected to represent the whole transducer. For example, in the case of Fig. 1, four networks can be used to represent the back mass, piezo stack, front mass and the horn. The behaviour of the transducer can then be analyzed from the obtained electrical model. Conventionally, the method is efficient in the transducer design, such as calculating the dimension based on the given frequency. But in the case of performance simulation or analysis, such as calculation of stress distribution, predication of power consumption, estimation of quality factor, etc, the application of this theory is limited. In the present work, we proposed an improved equivalent circuit theory, called Finite Electric Circuit Element Modelling (FECEM), which are convenient and efficient in both transducer design and analysis. In this method, the transducer will also be divided into multiple sections. But the section will be further divided into a number of subsections, called element. An element is represented using an electric circuit network. The connections of the networks constitute the electrical circuit model of the transducer. The number of elements can be flexibly selected based on the complexity of the structure and parameters to be solved. The advantages of this method are parameters such as mode shape, stress distribution and mechanical quality factor can be easily solved. In the following sections, the FECEM will be introduced using Langevin transducer as an example. For simplicity, the structure of horn is ignored. The transducer is considered to contain front mass, piezoelectric stack and back mass. The front and back masses are supposed to be made of the same material of aluminium. And the piezoelectric material is PZT.

3.1 Electrical circuit element

Naturally, the Langevin transducer mentioned above can be divided into front mass section, piezoelectric stack section and back mass section. The front and back mass sections are uniform cylinders. Therefore, they can be represented using the same element as shown in Fig. 2 (Lin, 1987).

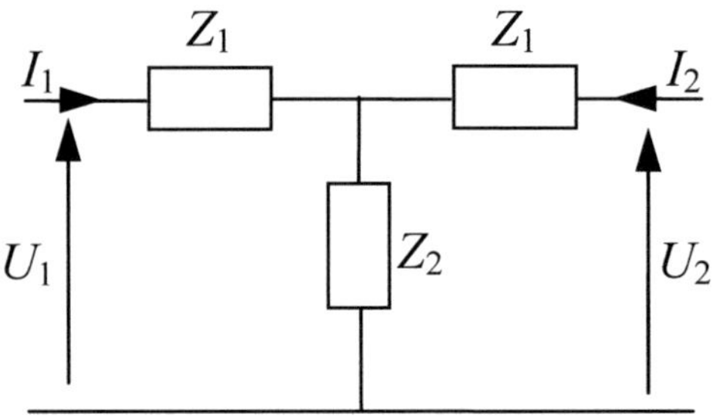

Fig. 2. Front and back mass elements

This is a two-port network, in which U_1, U_2 and I_1, I_2 are the voltages and currents at the input and output terminals. Physically, they mean the forces and velocities at the two ends of the vibrating element. Z_1 and Z_2 are impedances in the network. They can be expressed as

$$Z_1 = j\rho cS \tan\frac{kl}{2} \tag{1}$$

$$Z_2 = \frac{-j\rho cS}{\sin kl} \tag{2}$$

where ρ is density, c is sound speed, S is cross sectional area of the element, k is wave number and l is the length of the element. Based on Fig. 2, the two-port voltages can be expressed in terms of the two-port currents using the "Z" matrix (Balabanian & Bickart, 1981; Hayt & Kemmerly, 1993)

$$\begin{bmatrix} U_1 \\ U_2 \end{bmatrix} = \begin{bmatrix} Z_{11} & Z_{12} \\ Z_{21} & Z_{22} \end{bmatrix} \begin{bmatrix} I_1 \\ I_2 \end{bmatrix} \tag{3}$$

where

$$\begin{cases} Z_{11} = Z_1 + Z_2 \\ Z_{12} = Z_2 \\ Z_{21} = Z_2 \\ Z_{22} = Z_1 + Z_2 \end{cases} \tag{4}$$

This is actually the constitutive relations between the forces and velocities for the elements. This relation can also be represented using the following "chain" or "ABCD" matrix (Balabanian & Bickart, 1981; Hayt & Kemmerly, 1993)

$$\begin{bmatrix} U_1 \\ I_1 \end{bmatrix} = \begin{bmatrix} A & B \\ C & D \end{bmatrix} \begin{bmatrix} U_2 \\ -I_2 \end{bmatrix} \tag{5}$$

where

$$\begin{cases} A = (Z_1 + Z_2)/Z_2 \\ B = Z_1(Z_1 + 2Z_2)/Z_2 \\ C = 1/Z_2 \\ D = (Z_1 + Z_2)/Z_2 \end{cases} \tag{6}$$

The Z matrix and $ABCD$ matrix can be converted to each other based on the following Eqs. 7 and 8. This provides much convenience for different parameter calculations (Balabanian & Bickart, 1981; Hayt & Kemmerly, 1993).

$$\begin{bmatrix} A & B \\ C & D \end{bmatrix} = \begin{bmatrix} Z_{11}/Z_{21} & (Z_{11}Z_{22} - Z_{12}Z_{21})/Z_{21} \\ 1/Z_{21} & Z_{22}/Z_{21} \end{bmatrix} \tag{7}$$

$$\begin{bmatrix} Z_{11} & Z_{12} \\ Z_{21} & Z_{22} \end{bmatrix} = \begin{bmatrix} A/C & (AD - BC)/C \\ 1/C & D/C \end{bmatrix} \tag{8}$$

The piezoelectric stack comprises two parts, PZT and bolt. It can be represented using a three-port network element as shown in Fig. 3, in which V is the input electrical voltage, C_0

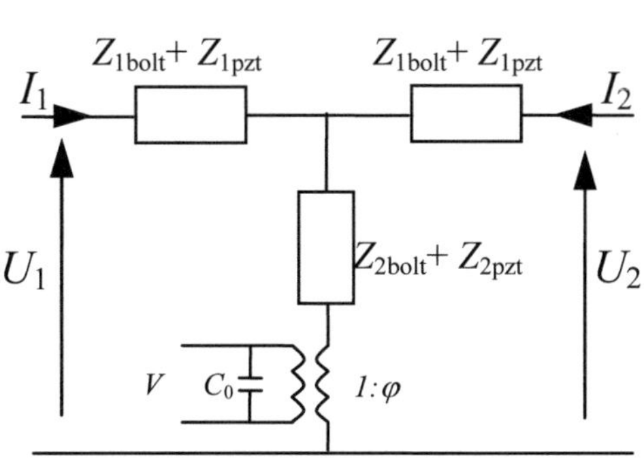

Fig. 3. Piezoelectric stack element (Lin, 2004b)

is the static capacitance, φ is the transformer ratio, Z_{1bolt}, Z_{2bolt}, Z_{1pzt} and Z_{2pzt} are the impedances from the bolt and PZT, respectively. Z_{1bolt} and Z_{2bolt} can also be expressed using Eqs. 1 and 2, respectively. The expressions for the rest parameters are shown below (Lin, 2004b; Sherrit et al., 1999a, 1999b).

$$Z_{1pzt} = j\rho c^D S \tan\frac{k^D l}{2} \tag{9}$$

$$Z_{2pzt} = \frac{-j\rho c^D S}{\sin k^D l} - \frac{\varphi^2}{j\omega C_0} \tag{10}$$

$$C_0 = \frac{S}{l\overline{\beta}_{33}} \tag{11}$$

$$\varphi = \frac{g_{33}S}{l s_{33}^D \overline{\beta}_{33}} \tag{12}$$

$$\overline{\beta}_{33} = \beta_{33}^T\left(1 + \frac{g_{33}^2}{s_{33}^D \beta_{33}^T}\right) \tag{13}$$

$$s_{33}^D = s_{33}^E\left(1 - {k_{33}}^2\right) \tag{14}$$

$$c^D = \sqrt{\frac{1}{\rho s_{33}^D}} \tag{15}$$

$$k^D = \frac{\omega}{c^D} \tag{16}$$

$$\beta_{33}^T = 1/\varepsilon_{33}^T \tag{17}$$

$$g_{33} = d_{33}\beta_{33}^T \tag{18}$$

$$k_{33} = \frac{d_{33}}{\sqrt{s_{33}^E \varepsilon_{33}^T}} \tag{19}$$

In the above expressions, d_{33} is piezoelectric coefficient, ε_{33}^T is dielectric constant, s_{33}^E is compliance coefficient. The constitutive relation for the piezoelectric stack element will hence be

$$\begin{bmatrix} U_1 - \varphi V \\ U_2 - \varphi V \end{bmatrix} = \begin{bmatrix} Z_{11} & Z_{12} \\ Z_{21} & Z_{22} \end{bmatrix} \begin{bmatrix} I_1 \\ I_2 \end{bmatrix} \tag{20}$$

where

$$\begin{cases} Z_{11} = Z_{1bolt} + Z_{1pzt} + Z_{2bolt} + Z_{2pzt} \\ Z_{12} = Z_{2bolt} + Z_{2pzt} \\ Z_{21} = Z_{2bolt} + Z_{2pzt} \\ Z_{22} = Z_{1bolt} + Z_{1pzt} + Z_{2bolt} + Z_{2pzt} \end{cases} \tag{21}$$

3.2 Finite electrical circuit element model

The vibrating elements are mechanically connected in series. Therefore the electrical model can be built as shown in Fig. 4 (Balabanian & Bickart, 1981; Hayt & Kemmerly, 1993).

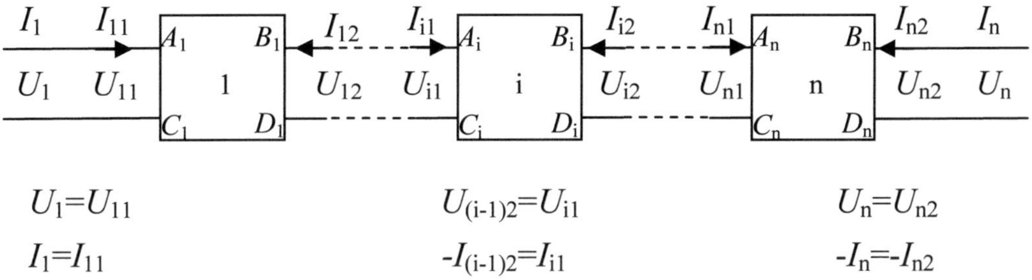

Fig. 4. Finite electrical circuit element model

The output of one element is the input of the adjacent one. Therefore, the following relation exists

$$\begin{bmatrix} U_1 \\ I_1 \end{bmatrix} = \begin{bmatrix} A_1 & B_1 \\ C_1 & D_1 \end{bmatrix} \bullet \cdots \bullet \begin{bmatrix} A_i & B_i \\ C_i & D_i \end{bmatrix} \bullet \cdots \bullet \begin{bmatrix} A_n & B_n \\ C_n & D_n \end{bmatrix} \begin{bmatrix} \vec{U}_n \\ -\vec{I}_n \end{bmatrix} = \begin{bmatrix} A & B \\ C & D \end{bmatrix} \begin{bmatrix} U_n \\ -I_n \end{bmatrix} \tag{22}$$

For the front mass, assuming there are n elements, the constitutive relation will be

$$\begin{bmatrix} U_1 \\ I_1 \end{bmatrix} = \begin{bmatrix} A_{front} & B_{front} \\ C_{front} & D_{front} \end{bmatrix} \begin{bmatrix} U_n \\ -I_n \end{bmatrix} \tag{23}$$

For the piezoelectric stack, assuming there are m elements, the following relation can be obtained

$$\begin{bmatrix} U_n - \varphi V \\ -I_n \end{bmatrix} = \begin{bmatrix} A_{pzt} & B_{pzt} \\ C_{pzt} & D_{pzt} \end{bmatrix} \begin{bmatrix} U_{n+m} - \varphi V \\ -I_{n+m} \end{bmatrix} \tag{24}$$

Rearranging, Eq. 24 turns to be

$$\begin{bmatrix} U_n \\ -I_n \end{bmatrix} = \begin{bmatrix} A_{pzt} & B_{pzt} \\ C_{pzt} & D_{pzt} \end{bmatrix} \begin{bmatrix} U_{n+m} \\ -I_{n+m} \end{bmatrix} + \begin{bmatrix} A_{pzt} - 1 & B_{pzt} \\ C_{pzt} & D_{pzt} - 1 \end{bmatrix} \begin{bmatrix} -\varphi V \\ 0 \end{bmatrix} \tag{25}$$

Similarly, for the back mass, supposing the number of elements is q, so there is

$$\begin{bmatrix} U_{n+m} \\ -I_{n+m} \end{bmatrix} = \begin{bmatrix} A_{back} & B_{back} \\ C_{back} & D_{back} \end{bmatrix} \begin{bmatrix} U_{n+m+q} \\ -I_{n+m+q} \end{bmatrix} \tag{26}$$

Combining Eqs. 23, 25 and 26, the relation for the whole system can be obtained as

$$\begin{bmatrix} U_1 \\ I_1 \end{bmatrix} = T_1 \begin{bmatrix} U_{n+m+q} \\ -I_{n+m+q} \end{bmatrix} + T_2 \begin{bmatrix} -\varphi V \\ 0 \end{bmatrix} \tag{27}$$

where

$$T_1 = \begin{bmatrix} A & B \\ C & D \end{bmatrix} = \begin{bmatrix} A_{front} & B_{front} \\ C_{front} & D_{front} \end{bmatrix} \begin{bmatrix} A_{pzt} & B_{pzt} \\ C_{pzt} & D_{pzt} \end{bmatrix} \begin{bmatrix} A_{back} & B_{back} \\ C_{back} & D_{back} \end{bmatrix} \tag{28}$$

$$T_2 = \begin{bmatrix} A' & B' \\ C' & D' \end{bmatrix} = \begin{bmatrix} A_{front} & B_{front} \\ C_{front} & D_{front} \end{bmatrix} \begin{bmatrix} A_{pzt} - 1 & B_{pzt} \\ C_{pzt} & D_{pzt} - 1 \end{bmatrix} \tag{29}$$

3.2.1 Modal analysis

Modal analysis is to find the resonant frequencies and the corresponding mode shapes of the transducer. The resonant frequencies can be found using the system Z matrix. Assuming the short circuit condition, i.e.,V=0, transform the *ABCD* matrix T_1 into the Z matrix Z_1 based on Eq. 7, and then set the determinant to zero

$$|Z_1| = 0 \tag{30}$$

The solutions of the equation are the resonant frequencies. The mode shape can be obtained based on the relation in Eq. 22. Assuming I_1=1, U_1=0, i.e., unit velocity and zero force at the free end, the velocity at each point can hence be solved. In another word, the mode shape can be found.

3.2.2 Harmonic analysis

The harmonic analysis is to find the performance of the transducer as a function of frequency. The parameters interested might include magnitude of velocity, magnitude of stress, power consumption, impedance characteristics, etc. This can be done based on the system relation Eq. 27. Assuming one mechanical port is loaded with a resistor R, the other mechanical port is short circuited, i.e., no force, and the electrical port is applied with 1 V, Eq. 27 will turn to

$$\begin{bmatrix} I_1 R \\ I_1 \end{bmatrix} = T_1 \begin{bmatrix} 0 \\ -I_{n+m+q} \end{bmatrix} + T_2 \begin{bmatrix} -\varphi V \\ 0 \end{bmatrix} \tag{31}$$

Solving the above equation, the velocity at the two ends (I_1 and I_{n+m+q}) can hence be obtained. The stress can then be estimated based on the mode shape, velocity obtained and the following relation (Silva, 1999)

$$T = E\varepsilon = \frac{E}{\omega}\frac{d\upsilon}{dx} \tag{32}$$

where E is Young's modus, υ is velocity and x is position. The stress is an important factor that affects heat generation and vibration amplitude (Li et al., 2007b, 2007c).

The input current to the transducer is solved using the following relation (Lin, 2004b)

$$I = V \cdot j\omega C_0 - \varphi(I_1 + I_2) \tag{33}$$

where I_1 and I_2 are velocities at the surface of the PZT. Solving the current, the electrical properties such as impedance and power consumption can hence be estimated accordingly.

3.2.3 Resistive load and mechanical quality factor

As seen in Section 3.2.2, before the harmonic analysis, an equivalent load R has to be assumed. Without it, the vibration magnitude of the transducer will go to infinity. In practice, the load can be the results of the external loadings or the internal losses. In the case of free vibration, the load is due to the internal mechanical loss, which is the function of the mechanical quality factor Q_m (Chen & Wu, 2002)

$$R = \frac{\omega M}{Q_m} \tag{34}$$

where M is the equivalent mass. To find M and Q_m, the following method is applied.

(a) Mechanical quality factor

The mechanical loss per unit volume is estimated using the following equation (Li et al., 2007b, 2007c)

$$W = \frac{1}{2}\omega s T^2 \tan\delta_m = \frac{1}{2} s T^2 \cdot \omega \cdot \frac{1}{Q_m} = w \cdot \omega \cdot \frac{1}{Q_m} \tag{35}$$

in which s is the compliance coefficient, T is the stress, $\tan\delta_m = 1/Q_m$ is the mechanical loss tangent, and $w = 1/2 sT^2$ is the elastic energy density. In Eq. 35, s and Q_m is solely material dependant. But the stress T not only depends on the materials, but also relies on the position x. Therefore, to calculate the total loss for a single material i, the following integration should be applied

$$W_i' = \int_0^l \omega w \frac{1}{Q_{mi}} A dx = \frac{1}{Q_{mi}} \omega \int_0^l w A dx = \omega W_i \frac{1}{Q_{mi}} \tag{36}$$

where A is the cross sectional area of the material, W_i is the elastic energy for material i. Assuming the transducer is made of n different materials, total mechanical loss of the transducer can hence be calculated as

$$W_{total}' = W_1' + W_2' + \cdots + W_n' = \omega \sum_{i=1}^{n} W_i \frac{1}{Q_{mi}} \tag{37}$$

On the other hand, the mechanical loss can also be expressed as

$$W'_{total} = \omega W_{total} \frac{1}{Q_m} \tag{38}$$

where $W_{total} = \sum_{1}^{n} W_i$ is the total elastic energy, and Q_m is the equivalent mechanical quality factor of the transducer. Equalizing Eq. 37 and 38, and rearranging, the following relation can finally be obtained

$$\frac{W_{total}}{Q_m} = \sum_{i=1}^{n} \frac{W_i}{Q_{mi}} \tag{39}$$

This equation means the equivalent quality factor Q_m can be estimated using the mechanical elastic energy and the individual material quality factor.

(b) Equivalent mass M

Because the piezoelectric transducer can be modelled using both lumped parameters and distributed parameters, the following relation exists (Li, 2004)

$$\frac{1}{2} M V_t^2 = \int_0^l \frac{1}{2} v_x^2 \rho A dx \tag{40}$$

where v_x is the velocity at position x, ρ is density, A is cross sectional area and V_t is the velocity at the tip. In Eq. 40, the left part is the kinetic energy of the transducer expressed in lumped parameters and right part in distributed parameters. From the derived mode shape mentioned earlier, v_x in Eq. 40 can be easily known. Therefore the equivalent mass can be conveniently solved.

3.2.4 Results for Langevin transducer

Table 1 and Fig. 5 compare the results between FECEM modelling and ANSYS 2D modelling (details in next section). It can be seen that the results are reasonably consistent. Therefore the FECEM theory is effective.

Model	Resonant frequency	Qm	velocity	Conductance G
ANSYS (2D)	65845	849	0.3	0.2
FECEM	67471	816	0.32	0.24
Difference	2%	4%	6%	20%

Table 1. Comparison of simulation results between ANSYS and FECEM theories

3.3 Concluding remarks

The FECEM theory is introduced in this section. This theory is useful for both transducer design and performance analysis. Based on this theory, the transducer can be divided into finite numbers of electrical cricuit elements, which are connected to form the electrical

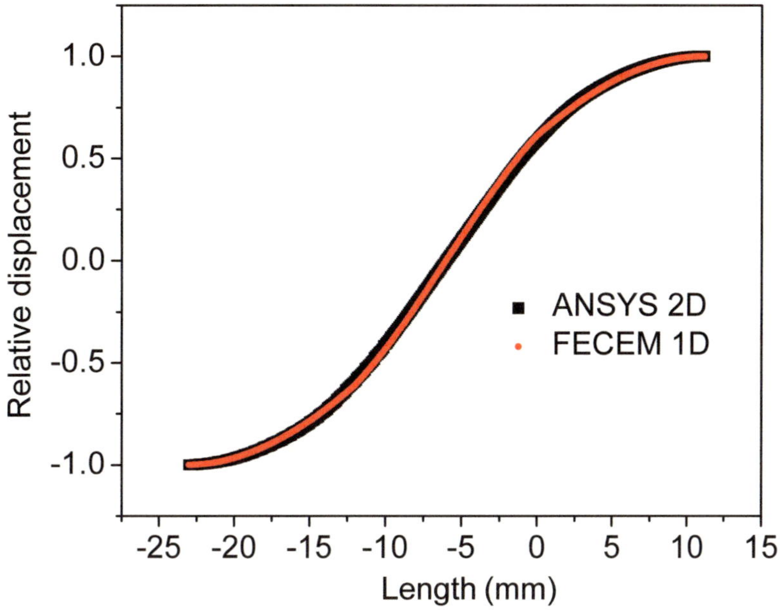

Fig. 5. Comparison of mode shape of Langevin transducer

circuit model. The mathematical respresention of the circuit model was also estibilished. Accordingly, the modal analysis and harmonic analysis can be achieved by solving the network equations. The mechanical quality factor and equivalent mass were derived, too. The FECEM was also compared with ANSYS 2D modelling. The results show reasonable accuracy.

4. ANSYS 2D modelling

The piezoelectric transducer can also be analyzed using ANSYS 2D simulation (ANSYS, Inc., 2010; Li et al., 2008, 2009, 2010). One of the advantages of this method is that the acoustic field with various boundary conditions can be coupled to the piezoelectric transducer. Therefore, the model is useful to analyze the response of the transducer under different loading conditions. It is also essential for the study of the sound pressure distributions at different boundary conditions.

Fig. 6 shows the transducer-fluid coupled model, in which the 2D model revolving about the axial of rotation will produce the 3D structure of the system. The liquid medium is water which directly contacts the mechanical structure of the piezoelectric transducer at the structure-liquid interface. For the piezoelectric transducers, the axisymmetric structure element Plane42 for metallic parts and couple field element Plane13 for PZT materials were applied. The Plane42 element has degrees of freedom UX and UY, and Plane13 element has degrees of freedom VOLT, UX and UY. For liquid medium, acoustic element Fluid29 was applied. In the acoustic domain, it has degrees of freedom PRES. And at the boundary and interface, it possesses degrees of freedom UX, UY and PRES. The boundary condition can be controlled by the absorption coefficient MU. In this case, the horizontal boundary was set "hard" and curved boundary is set "absorbing", corresponding to MU=0 and MU=1, respectively. Fig. 6 (b) shows the meshed model of the system. After meshing, an electrical load of 1 V was applied to the PZT stack. To be noted, the transducer analysis and the acoustic analysis can be modelled separately, which might be more efficient for certain problems.

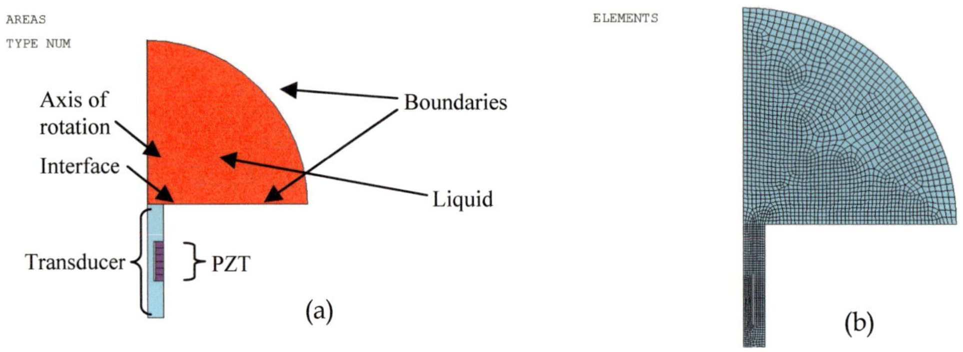

Fig. 6. 2D modelling of the transducer-fluid coupled system, (a) model and (b) meshed model

Fig. 7 shows the harmonic analysis results of the transducer displacement at the resonant frequency. The colour represents the magnitude of the displacement in the unit of meter. ANSYS provides the harmonic displacement with the real and imaginary part. Because at resonance, the electrical excitation and displacement response has 90 deg phase difference, the imaginary part dominates. It is also noticed that the resonant frequency is 64510 Hz in the figure, which is different from 65845 Hz in Table 1. This is because the liquid loading changes the resonant frequency of the transducer.

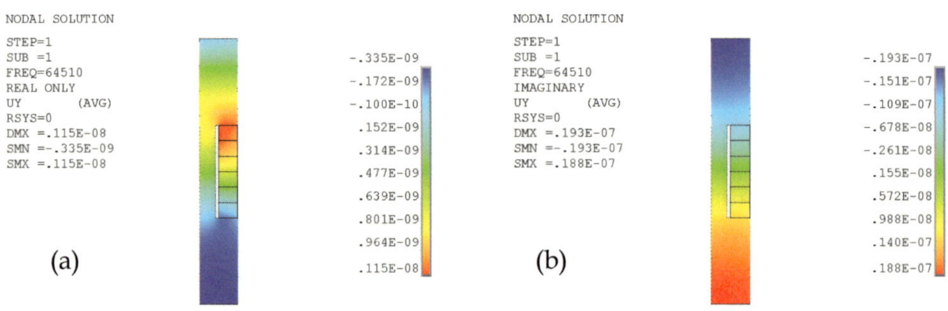

Fig. 7. Real and imaginary part of the transducer displacement, (a) real, (b) imaginary

Fig. 8 shows the acoustic pressure distribution in the acoustic field. The colour represents the magnitude of the pressure wave in the unit of dB. In this case, the transducer-fluid interface is the pressure wave radiation surface. The horizontal hard boundary resembles an infinite baffle. A travelling pressure wave is therefore generated at the transducer radiation surface. It then propagates in the half hemisphere liquid domain. And finally it is fully absorbed at the curved absorbing boundary. The maximum pressure is located at the centre of the transducer surface. It decreases as the distance from the transducer surface increases. The magnitude and distribution of the pressure wave are sensitive to the boundary conditions. Therefore the boundary conditions should be optimized accordingly based on the applications.

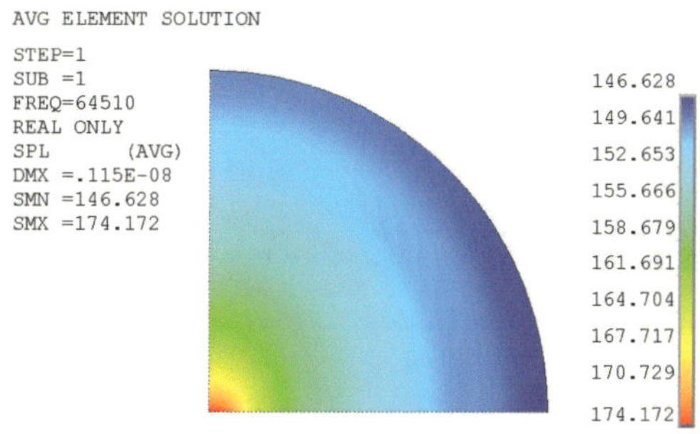

Fig. 8. Sound pressure pattern in the liquid domain

5. Thrombolysis transducer

Thrombolysis transducer is a device used to deal with the vascular thrombotic occlusive disease, which is a major cause of morbidity and mortality in the developed world. The development of a blood clot or thrombus in a blood vessel compromises distal blood flow and is the usual cause of a heart attack or stroke. Established treatment is the urgent removal or dissipation of the occluding thrombus. This is achieved with the use of a simple aspiration catheter, mechanical thrombectomy, or pharmacological agents such as thrombolytic drugs (Atar et al., 1999; Brosh et al., 1998; Janis et al., 2000; Ma et al., 2008; Siegel & Luo, 2008). Ultrasonic emulsification of the blood clot is another technique for thrombolysis. This is achieved by acoustic cavitation and mechanical fragmentation (Bond & Cimino, 1996; Chan et al., 1986). Compared with conventional mechanical thrombectomy techniques ultrasonic thrombolysis exhibits the advantage of inherent tissue selectivity (Rosenschein et al., 1994; Tschepe et al., 1994). This is because thrombus is highly susceptible to ultrasonic cavitational emulsification, while the arterial walls, which are lined with cavitation-resistant matrix of collagen and elastin, are not. Ultrasound energy has also been shown to improve myocardial reperfusion in the presence of coronary occlusion (Siegel et al., 2004). The system of ultrasonic thrombolysis generally contains a generator, a transducer and a catheter. The tip of the transducer will go through the catheter lumen to the blood clot. Then the generator will provide control signal and power to the transducer. The vibration of the tip will emulsify the blood clot. In the following sections, the thrombolysis transducer developed in NTU (Nanyang Technological University) will be introduced.

5.1 Structure

Fig. 9 shows the structure of the piezoelectric thrombolysis transducer designed in the present work. The transducer consists of five parts. The first part is the end cap. It serves to prestress the adjacent PZT stack and also adjusts the mechanical impedance applied to the stack. The second part is the PZT stack, which is clamped between the end cap and the horn. It is the most crucial part of the transducer, where the vibration is generated. The third part of the transducer is the horn, which functions to magnify the displacement produced by the PZT stack. The fourth part is a long and thin transmission wire, which should be flexible but sufficiently stiff for energy transmission. The last part is a distal vibration tip that consists of a ball or a short cylinder with an enlarged diameter (~ 1.5 mm) compared to the connecting transmission wire (~ 0.5 mm). The enlarged diameter increases acoustic power emission to the surrounding liquid and blood clot. The vibration produced by the PZT stack is transmitted through the horn and then the transmission wire to the distal vibration tip. The acoustic energy emitted from the tip is finally used to emulsify the clot.

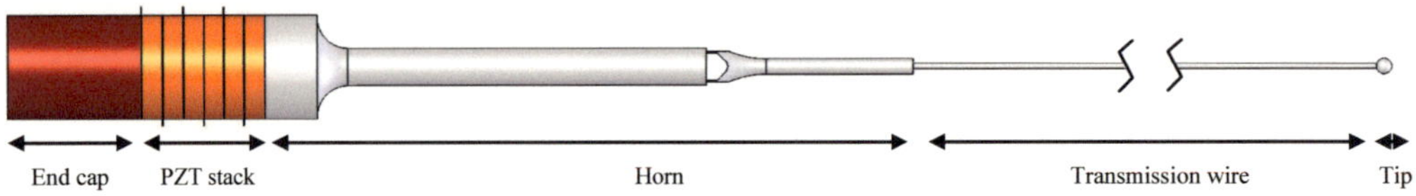

Fig. 9. Schematic illustration of the piezoelectric thrombolysis device

The transducer operates at ~26.7 kHz longitudinal vibration mode. This is a low ultrasonic frequency aiming at reducing the heat generation (Francis, 2001; Li et al., 2007b; Siegel et al., 2000). The diameter of the PZT stack is 10 mm. Maximum input power is 20 W. The length of transmission wire is 1 m, which is made of a high strength material, Ti-6Al-4V, for achieving a high vibration velocity (Li et al., 2007b; Mason & Wehr, 1970; Muhlen, 1990). For practical operations, due to the long length of the transmission wire, the bending mode could be excited, which would increase loss and decrease the efficiency of the device. To avoid this, the transmission wire should be coaxial with the horn.

5.2 FECEM modeling

Because of the high aspect ratio of the device (length/diameter), it is very advantageous to model the device using the FECEM theory. Fig. 10 shows the mode shape of the transducer. It can be seen that along the axial direction, the displacement generated by the PZT stack is first amplified by the horn, then further amplified by the transmission wire. As a result, the tip has a much larger displacement than the PZT stack. Also considering the smaller area of the vibration tip to the PZT stack, it can be said that the energy produced by the PZT stack has been focused to the tip through this design. Therefore, the tip will work effectively for the blood clot emulsification.

5.3 ANSYS 2D acoustic modeling

During practical operation, the vibration tip will be surrounded by liquid and produces an acoustic field. Fig. 11 shows the simulation of the acoustic field generated by an OD 1.5 mm and length 3 mm vibration tip. The tip is connected to an OD 0.5 mm transmission wire, and the vibration frequency of the tip is 30 kHz. The simulation was carried out using ANSYS 2D acoustic analysis. Fig. 11 shows that the maximum ultrasonic pressure is located at the top and bottom surfaces of the vibration tip, which is normal to the displacement direction. This indicates that emulsification should be most effective at these locations. The radiation area multiplied by the normal surface velocity of the tip is known as the source strength (International standard, IEC 1998; Li et al., 2009). The ultrasonic pressure amplitude is

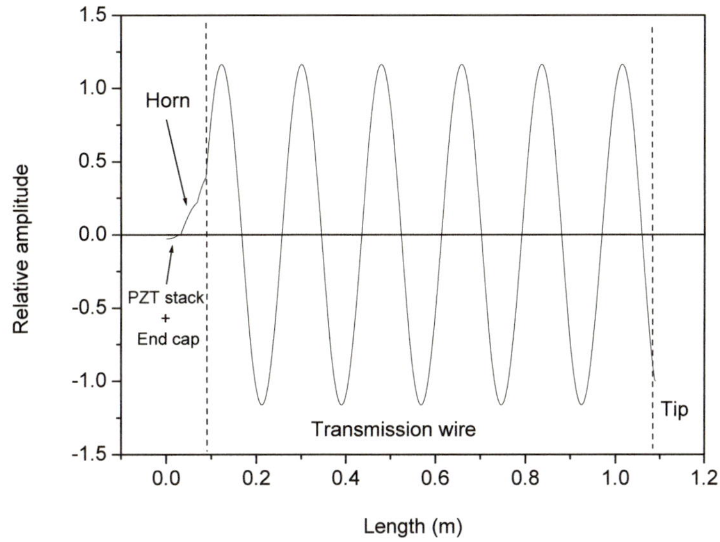

Fig. 10. Vibration amplitude distribution along the length of the transducer

proportional to the source strength (International standard, IEC 1998; Li et al., 2009). Therefore, in the present work, a horn is applied to amplify the vibration velocity from the PZT crystal. And a ball or a short cylinder tip is attached at the distal end of the transmission wire to enhance the radiation area.

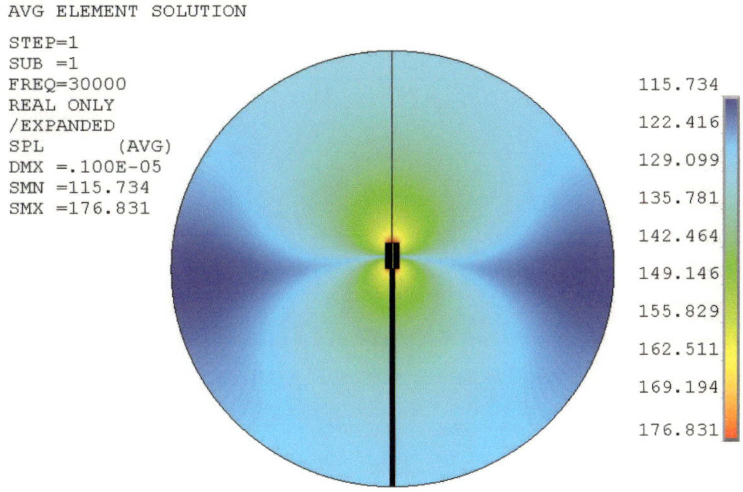

Fig. 11. Acoustic pressure pattern generated by the tip in the liquid

5.4 Acoustic characterization

The acoustic properties of the transducer were characterized qualitatively. As the acoustic pressure becomes larger and larger, two phenomena might be observed around the tip of the transducer, i.e., cavitation and streaming (Abramov, 1994, 1998; Young, 1999). Cavitation is the generation and burst of bubbles in the liquid due to the high amplitude of the acoustic pressure. Along with the burst of the bubbles is the high intensity shock wave and impinging of the liquid, which is normally very strong and even sufficient to break a hard surface. Fig. 12 (a) shows the cavitation bubble clusters generated at the vibration tip in the silicon oil. The cluster is usually generated at the center of the surface and then flows away along the acoustic axis. When the distance becomes larger from the surface, the bubbles might be agglomerating and floating upwards due to the buoyancy force. The cavitation threshold of the water is larger than silicon oil and is also frequency, temperature and static pressure dependant (Abramov, 1994, 1998; Young, 1999). Generally in water, only when the acoustic field is very strong, visible bubble clusters can be observed. The generation of the bubble cluster at the tip surface also proves that the highest intensity of acoustic energy is around the tip. This is consistent with the earlier theoretical analysis.

The second important phenomenon is acoustic streaming, which is the flow of the liquid as the result of high and nonlinear acoustic pressure field and the generation of bubble clusters (Abramov, 1994). The flow pattern is usually that the flow is along the acoustic axis and outwards, the surrounding liquid will flow to the tip and compensate the outflows. The streaming effects have both positive and negative effect in application of ultrasonic thrombolysis. For the positive aspect, the streaming will expedite the ablation of the blood clot, especially when the fibrinolytic agents are present (Atar et al., 1999). The negative effect is that the streaming is inclined to push away the blood clot, which increases the difficulty to control, especially, the floating clot. Fig. 12 (b) is the demonstration of acoustic streaming, which pushes the water away from the tip surface. Because the demonstration is near the water surface, the droplets jumped into the air, resulting in the phenomenon so called atomization (NII et al., 2006).

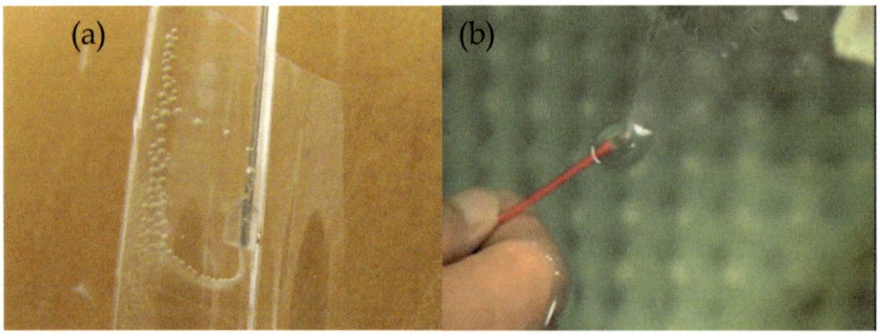

Fig. 12. Acoustic cavitation (a) and acoustic streaming (b)

5.5 Blood clot emulsification

The transducer was tested to emulsify the blood clot in an anechoic tank. The tank was filled with water and lined with sound absorption materials both at the walls and the bottom. The dimensions of the tank are $0.6 \times 0.6 \times 1.3$ m^3. A holder made of natural latex of 30 μm thick was used to contain the blood clot. The clot was prepared by naturally coagulating fresh rabbit blood overnight at 6 ^{o}C. During operation, the tip of the transducer was pointed at the clot surface as shown in Fig. 13. The blood clot was immediately emulsified when the power was provided. The whole procedure documents rapid clot lysis (~ 750 mg/min) and confirm the effectiveness of the transducer in thrombolysis.

Fig. 13. Emulsification of the blood clot by the transducer tip

5.6 Concluding remarks

The piezoelectric thrombolysis transducer consists of an end cap, a PZT stack, a horn, a transmission wire and a vibration tip. The transducers vibrate longitudinally and generate maximum acoustic pressure at the tip. The acoustic pressure induces effects such as cavitation and streaming at the tip. Blood clot can be effectively emulsified using the designed transducer.

6. Acoustic pump

The piezoelectric pump has various designs (Laser & Santiago, 2004; Luong & Nguyen, 2010; Yeo et al., 2008). According to frequency, the pumping mechanism will be different. At the sonic low frequency range, the diaphragm pump is a popular design, which utilizes the reciprocating movement of piezo transducer to displace the liquid. However, as frequency increases, the diaphragm mechanism becomes more and more difficult to achieve. One of the important reasons is the response mismatch between the diaphragm and valves (Hu et al., 2004; Zhang et al., 2003). However, there are other pumping mechanisms. At the ultrasonic and megasonic frequency range, the acoustic effect could be used (Chen & Lal, 2006; Frampton et al., 2004; Hasegawa et al., 2008; Koyama, et al., 2010; Li et al., 2010; Zhang et al., 2010). It has been well known that the intense and nonlinear ultrasound could induce effect such as cavitation and streaming, which are able to generate the flow of the liquid (Abramov, 1994; Li et al., 2010; Raton et al., 2007). This work will report a piezoelectric acoustic pump based on these acoustic effects.

6.1 Structure of the pump

Fig. 14 shows the structure of the piezoelectric pump designed in NTU. The pump comprises a piezoelectric transducer, a reflecting condenser and a casing. The structure of the transducer is similar as the thrombolysis transducer, but without transmission wire. There is a through hole inside the transducer working as the flow path. The transducer works approximately at 30 kHz. The reflecting condenser has a "⊥" shape, which focused the acoustic energy at its corner. The casing has a cylindrical shape with inlet on the wall. The casing, condenser and horn surface form the boundaries of the acoustic field. Because the pump mechanism is based on the acoustic effect, the pressure magnitude and pattern inside the field will affect the performance of the pump significantly. The distribution of sound pressure level inside the casing is therefore simulated in the next section.

6.2 Modeling and simulation

Fig. 15 is the simulation of the pressure pattern inside the pump casing. The casing was simplified as a cylinder in this 2D axisymmetric simulation. The length of the cylinder is one wavelength. The graph shows the right half piece of the model. The horn tip generates 1 μm displacement at 30 kHz. The colour represents the pressure magnitude in the unit of dB. Two boundary conditions are compared. Fig. 15 (a) is the absorbing condition. As mentioned earlier, a travelling wave will be generated at the tip. It propagates inside the casing and finally is absorbed at the casing boundary. The maximum pressure is located at the horn tip. The magnitude is approximately 187 dB. Fig. 15 (b) shows the hard or reflecting boundary condition. In this case, a standing wave will be established inside the casing. The maximum pressure is still at the horn tip. But the magnitude is increased to 255 dB, which is much larger than that of the absorbing boundary condition. This indicates that the reflecting boundary will be more effective than the absorbing boundary to focus the energy or increase the pressure magnitude. In practice, the reflecting boundary could be achieved by using materials with high acoustic characteristic impedance such as metals (Feng, 1999).

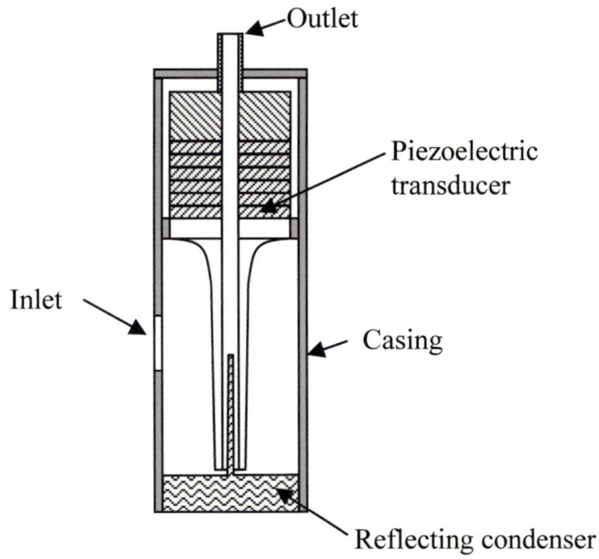

Fig. 14. The schematic structure of the piezoelectric acoustic pump designed in NTU

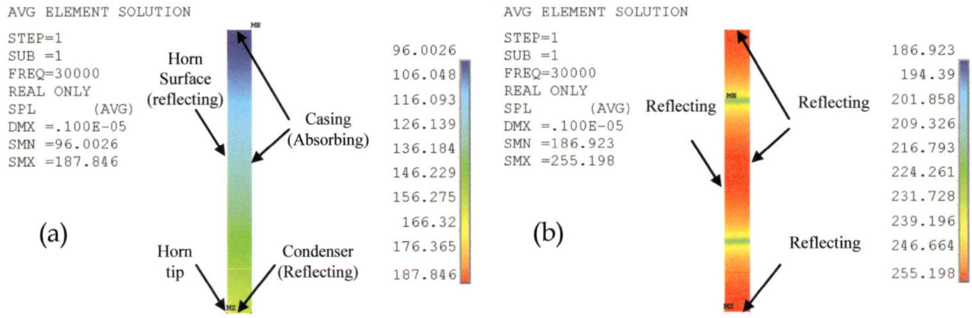

Fig. 15. The pressure pattern inside casing, (a) Absorbing boundary, (b) Reflecting boundary

6.3 Performance test

Fig. 16 shows the performance of the pump. From Fig. 16 (a), it can be seen that the maximum flow rate is about 10 mL/min. And maximum pressure head is about 250 kPa. The maximum pressure head is voltage dependant. As voltage increases, the pressure head can be further increased as shown in Fig. 16 (b). Compared with the regular piezoelectric diaphragm pump whose pressure head is mostly in the magnitude of ~50 kPa (Laser & Santiago, 2004; Luong & Nguyen, 2010), the advantage of the acoustic pump is the high pressure head. The reason for this is that the acoustic cavitational effect contains large energy, which is able to generate a high force to overcome the back pressure.

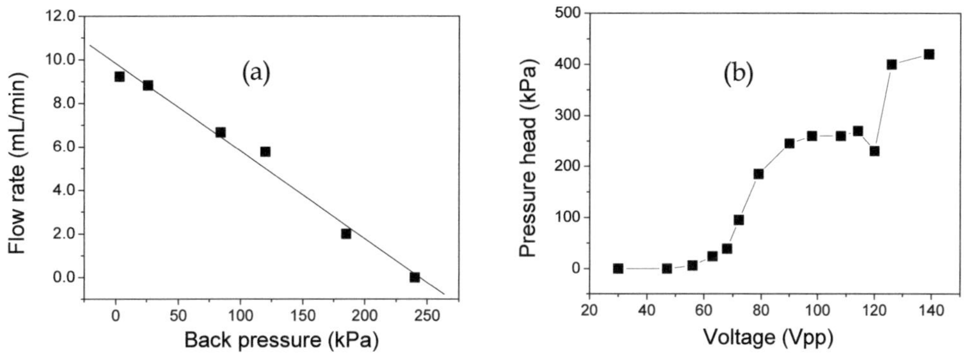

Fig. 16. Performance of the acoustic pump, (a) flow rate *vs* back pressure, (b) pressure head *vs* voltage

6.4 Concluding remarks

The acoustic pump is based on the nonlinear acoustic effect such as cavitation and streaming. The cavitation effect contributes significantly on the high pressure feature of the pump. The advantage of the piezoelectric acoustic pump designed in NTU is simple structure, less oscillation and high pressure head.

7. Conclusions

The general structure of the horn-type transducer is the Langevin transducer with a horn. The advantage of such device is that both large amplitude and high power (density) can be obtained simultaneously. FECEM theory can be used for the development of the piezoelectric transducer. This method is based on two-port network theory. Parameters like dimension, frequency, stress, velocity, quality factor, impedance, etc, can all be solved using this method. It also shows consistent results with ANSYS 2D analysis. The ANSYS 2D analysis is effective for the piezoelectric structure, acoustic field, or structure-acoustic coupled field analysis. It is also important in the development of piezoelectric devices. Two applications of horn-type transducer were also introduced in this work. The thrombolysis transducer is used for blood clot emulsification. Due to high aspect ratio of the structure, it is also a good example to be analyzed using the FECEM theory. The acoustic pump, on the other hand, possesses a simple structure for pumping applications. The working principle is nonlinear acoustic effects of streaming and cavitation. The major advantage of the pump is the high pressure head. It might be used for sonochemistry or microfluidics.

8. References

Abramov, O.V. (1994). *Ultrasound in Liquid and Solid Metals*, CRC Press, ISBN 0849393558, Boca Raton

Abramov, O.V. (1998). *High-Intensity Ultrasonics Theory and Industrial Applications*, Cordon and Breach Science Publishers, ISBN 9056990411, Russia

Accoto, D.; Carrozza M.C. & Dario P. (2000). Modelling of Micropumps Using Unimorph Piezoelectric Actuator and Ball Valves, *J. Micromech. Microeng.*, Vol.10, No. 2, (June 2000), pp. 277-281, ISSN 0960-1317

ANSYS, Inc. (2010) *ANSYS Fluid Analysis Guide and Coupled-Field Analysis Guide*, Release 13

APC international, Ltd. (2002). *Piezoelectric Ceramics: Principles and Applications*, APC International, Ltd., ISBN 0971874409, Mackeyville

Atar, S.; Luo, H.; Nagai, T. and Siegel, R.J. (1999). Ultrasonic Thrombolysis: Catheter-Delivered and Transcutaneous Applications. *European Journal of Ultrasound*, Vol.9, No.1 (March 1999), pp. 39–54, ISSN 09298266

Bai, M.R.; Chen R.L.; Chuang C.Y.; Yu C.S. & Hsieh H.L. (2007). Optimal Design of Resonant Piezoelectric Buzzer from a Perspective of Vibration-Absorber Theory, *J. Acoust. Soc. Am.*, Vol.122, No.3, (September 2007), pp. 1568-1580, ISSN 0001-4966

Balabanian, N. & Bickart, T.A. (1981). *Linear Network Theory Analysis, Properties, Design and Synthesis*, Matrix publisher, Inc., ISBN 0-916460-10-X, Beaverton

Bond, L.J. & Cimino W.W. (1996). Physics of Ultrasonic Surgery Using Tissue Fragmentation, *Ultrasonics*, Vol. 34, No. 2-5, (June 1996), pp. 579-585, ISSN 0041-624X

Brosh, D.; Miller, H.I.; Herz, I.; Laniado, S. & Rosenschein, U. (1998) Ultrasound Angioplasty: an Update Review, *International Journal of Cardiovascular Interventions*, Vol.1, No. 1, (January 1998), pp.11-18, ISSN 1462-8848

Chan, K.; Watmough, D.J.; Hope D.T. & MOIR, K. (1986). A New Motor-Driven Surgical Probe and its In Vitro Comparison with the Cavitron Ultrasonic Surgical Aspirator, *Ultrasound in Medicine & Biology*, Vol.12, No.4, (April 1986), pp.279-283, ISSN 0301-5629

Chen, X. & Lal A. (2006). Theoretical and Experimental Study of a High Flow Rate Ultrasonic Horn Pump, *2006 IEEE Ultrasonics Symposium*, pp.2409-2412, ISBN 1-4244-0201-8, Vancouver, BC, October 2-6, 2006

Chen, Y.C. & Wu, S. (2002). A Design Approach of Tonpiltz Transducer, *Jpn. J. Appl. Phys.*, Vol. 41, No.6A, (June 2002), pp. 3866-3877, ISSN 00214922

Chen, Y.C.; Wu, S. & Chen P.C. (2004). The Impedance-Matching Design and Simulation on High Power Electro-acoustical Transducer, *Sensors and Actuators A*, Vol.115, No. 1, (September 2004), pp.38-45, ISSN 0924-4247

Chu, W.P.; Li, H.L.; Chan, L.W.; Ng, M.W. & Liu C.K. (2002). Smart Ultrasonic Transducer for Wire-Bonding Applications, *Materials Chemistry and Physics*, Vol. 75, No.1-3 (April 2002), pp.95-100, ISSN 0254-0584

Feng, R. (1999). Ultrasonics Handbook, Nanjing University, ISBN 7-305-03354-5, Nanjing

Frampton, K.D.; Minor, K. & Martin, S. (2004). Acoustic Streaming in Micro-Scale Cylindrical Channels, *Applied Acoustics*, Vol. 65, No.11, (November 2004), pp.1121-1129, ISSN 0003-682X

Francis, C.W. (2001). Ultrasound-Enhanced Thrombolysis. *Echocardiography*, Vol. 18, No. 3, (April 2001), pp. 239-246, ISSN 0742-2822

Hasegawa, T.; Koyama D.; Nakamura, K. & Ueha, S. (2008). Modeling and Performance Evaluation of an Ultrasonic Suction Pump, *Japanese Journal of Applied Physics*, Vol.47, No. 5, (May 2008), pp.4248-4252, ISSN 00214922

Hayt, W.H. & Kemmerly J.E. (1993). *Engineering Circuit Analysis* (Sixth edition), McGRAW-HILL, Inc., ISBN 0-07-228364-5, New York

Hongoh, M.; Yoshikuni, M.; Miura, H.; Miyamoto, R.; Ueoka T. & Tsujino, J. (2004) Configuration of a 30-mm-Diameter 94 kHz Ultrasonic Longitudinal Vibration System for Plastic Welding, *Japanese Journal of Applied Physics*, Vol.43, No.5b, (May 2004), pp.2896-2900, ISSN 0021-4922

Horita, R.E. (1967). Free-Flooding Unidirectional Resonators for Deep-Ocean Transducers, *The Journal of the Acoustical Society of America*, Vol.41, No.1, (January 1967), pp.158-166

Hu, M.; Du, H.J.; Ling, S.F.; Fu, Y.Q.; Chen, Q.F.; Chow, L. & Li, B. (2004). A Silicon-on-Insulator Based Micro Check Vavle, *J. Micromech. Microeng.*, Vol.14, No.3, (March 2004) pp.382-387, ISSN 0960-1317

International standard, IEC 61847, (1998-01). *Ultrasonics-Surgical Systems-Measurement and Declaration of the Basic Output Characteristics*

Janis, A.D.; Buckely, L.A. & Gregory, K.W. (2000). Laser Thrombolysis in an In-Vitro Model, *Proc. SPIE - The International Society for Optical Engineering*, Vol.3907, pp.582-599, ISSN 0277-786X, San Jose, CA, USA, January 22-25, 2000

Kapila, V.; Deymier, P.A.; Shende, H.; Pandit, V.; Raghavan S. & Eschbach, F.O. (2006). Megasonic Cleaning, Cavitation, and Substrate Damage: an Atomistic Approach, *Proc. Of SPIE - The International Society for Optical Engineering*, Vol.6283, pp.628324-1- 12, ISSN 0277-786X, Yokohama, Japan, April 18, 2006

Khmelev, V.N.; Tchyganok,S.N.; Barsukov, R.V. & Lebedev, A.N. (2005). Design and Efficiency Analysis of Half-Wave Piezoelectric Ultrasonic Oscillatory Systems, *6th International Siberian workshop and tutorial EDM'2005, Session II*, pp.82-85, ISBN 5778204914, Erlagol, July 1-5, 2005

Koyama, D.; Wada, Y.: Nakamura, K.; Nishikawa, M.; Nakagawa, T. & Kihara, H. (2010). An Ultrasonic Air Pump Using an Acoustic Travelling Wave Along a Small Air Gap, *IEEE Transactions on Ultrasonics, Ferroelectrics, and Frequency Control*, Vol.57, No.1, (January 2010), pp.253-261, ISSN 0885-3010

Laser, D.J. & Santiago J.G. (2004). A Review of Micropumps, *J. Micromech. Microeng.*, Vol.14, No.6, (June 2004), pp.R35-R64, ISSN 0960-1317

Lee, D.R. & Loh, B.G. (2007). Smart Cooling Technology Utilizing Acoustic Streaming, *IEEE Transactions on Components and Packaging Technologies*, Vol.30, No.4, (December 2007), pp.691-699, ISSN 1521-3331

Li, Tao (2004). *Development of Piezoelectric Tubes for Micromotor*, PhD thesis, Nanyang Technological University

Li, Tao; Chen, Y.H.; Ma, J. & Boey F.Y.C. (2007a). Metal-PZT Composite Piezoelectric Transducers and Ultrasonic Motors, *Key Engineering Materials*, Vol.334-335, (2007) pp. 1073-1076, ISSN 1013-9826

Li, Tao; Chen, Y.H. & Ma, J. (2007b). Frequency Dependence of Piezoelectric Vibration Velocity, *Sensors and Actuators A*, Vol.138, No.2, (August 2007) pp.404-410, ISSN 0924-4247

Li, Tao; Chen, Y.H.; Boey, F.Y.C. & Ma, J. (2007c). High Amplitude Vibration of Piezoelectric Bending Actuators, *J. Electroceram*, Vol.18, No.3-4, (August 2007), pp.231-242, ISSN 13853449

Li, Tao; Chen, Y.H. & Ma, J. (2008). Development of Miniaturized Piezoelectric Multimode Transducer for Projection Purpose, *Advanced Materials Research*, Vol.47-50, (2008) pp.61-64, ISSN 10226680

Li, Tao; Chen, Y.H. & Ma, J. (2009). Development of a Miniaturized Piezoelectric Ultrasonic Transducer, *IEEE Transactions on Ultrasonics, Ferroelectrics, and Frequency Control*, Vol.56, No.3, (March 2009), pp. 649-659,ISSN 0885-3010

Li, Tao; Chen, Y.H; Lew F.L. & Ma J. (2010). Design, Characterization, and Analysis of a Miniaturized Piezoelectric Transducer, *Materials and Manufacturing Processes*, Vol.25, No.4, (April 2010), pp.221-226, ISSN 10426914

Lin, S.Y. (1995). Study on the Multifrequency Langevin Ultrasonic Transducer, *Ultrasonics*, Vol. 33, No.6, (November 1995), pp.445-448, ISSN 0041-624X

Lin, S.Y. (2004a). Effect of Electric Load Impedances on the Performance of Sandwich Piezoelectric Transducers, *IEEE Trans. Ultrason., Ferroelec., Freq. Contr.*, Vol.5, No.10 (October 2004), pp.1280-1286, ISSN 0885-3010

Lin, S.Y. (2004b). *Ultrasonic Transducer Principle and Design*, Science Press, ISBN 7-03-013419-2, Beijing

Lin, Z.M. (1987). *Ultrasonic Horn Principle and Design*, Science press, ISBN 7-03-000008-0, Beijing

Luong, T.D. & Nguyen N.T. (2010). Surface Acoustic Wave Driven Microfluidics - A Review, *Micro and Nanosystems*, Vol.2, No.3, (2010) pp.217-225, ISSN 18764029

Ma, J., Low F.H.A. & Boey Y.C.F. (2010) *Micro-Emulsifier for Arterial Thrombus Removal*, PCT/SG2008/000323, WO 2010/027325 A1

Mason, W.P. (1964). *Physical Acoustics Principles and Methods, Volume I - Part A*, Academic Press, ISBN 0124779018, New York

Mason, W.P. & Wehr J. (1970). Internal Friction and Ultrasonic Yield Stress of the Alloy 90 Ti 6 Al 4 V, *J. Phys. Chem. Solids*, Vol.31, No.8, (August 1970), pp.1925-1933, ISSN 0022-3697

Mattiat, O.E. (1971), *Ultrasonic Transducer Materials*, Plenum Press, ISBN 0306305011, New York

Medis, P.S. & Henderson H.T. (2005). Micromachining Using Ultrasonic Impact Grinding, *J. Micromech. Microeng.*, Vol.15, No. 8, (August 2005), pp.1556-1559, ISSN 0960-1317

Miclea, C.; Tanasoiu, C.; Iuga, A.; Spanulescu, I.; Miclea, C.F.; Plavitu, C.; Amarande, L.; Cioangher, M.; Trupina, L.; Miclea, C.T. & Tanasoiu, T. (2008). A High Performance PZT Type Material Used as Sensor for an Audio High Frequency Piezoelectric Siren, *2008 International Semiconductor Conference*, pp.185-188, ISBN 978-1-4244-2004-9, Sinaia, Romania, October 13-15, 2008

Minchenko, H. (1969). High-Power Piezoelectric Transducer Design, *IEEE Transactions on Sonics and Ultrasonics*, Vol. SU-16, No. 3, (July 1969), pp.126-131, ISSN 0018-9537

Muhlen, S.S. (1990). Design of an Optimized High-Power Ultrasonic Transducer, *1990 IEEE Ultrasonics Symposium*, pp.1631-1634, Honolulu, HI , USA, December 04-07, 1990

NII, S.; Matsuura, K.; Fukazu, T.; Toki, M. & Kawaizumi F. (2006). A Novel Method to Separate Organic Compounds through Ultrasonic Atomization, *Chemical*

Engineering Research and Design, Vol.84, No.5A, (May 2006), pp.412-41, ISSN 02638762

Prokic, M. (2004). *Piezoelectric Transducers Modeling and Characterization*, MPI, Switzerland

Radmanovic, M.D. & Mancic D.D. (2004). *Design and Modelling of the Power Ultrasonic Transducers*, MPI, ISBN 86-80135-87-9, Switzerland

Rosenschein, U.; Frimerman, A.; Laniado S. & Miller, H.I. (1994). Study of the Mechanism of Ultrasound Angioplasty from Human Thrombi and Bovine Aorta, *Am. J. Cardiol.*, Vol.74, No.12, (December 1994), pp.1263-1266, ISSN 0002-9149

Sherrit, S.; Dolgin, B.P.; Bar-Cohen, Y.; Pal, D.; Kroh J. & Peterson, T. (1999a). Modeling of Horns for Sonic/Ultrasonic Applications, *1999 IEEE Ultrasonics Symposium*, pp.647-651, ISBN 0-7803-5722-1, Caesars Tahoe, NV, USA, October 17-20, 1999

Sherrit, S.; Leary, S.P.; Dolgin B.P. & Bar-Cohen Y. (1999b). Comparison of the Mason and KLM Equivalent Circuits for Piezoelectric Resonators in the Thickness Mode, *1999 IEEE Ultrasonics Symposium*, pp.921-926, ISBN 0-7803-5722-1, Caesars Tahoe, NV, USA, October 17-20, 1999.

Siegel, J.; Atar, S.; Fishbein, M.C.; Brasch, A.V.; Peterson, T.M.; Nagai, T.; Pal, D.; Nishioka, T.; Chae, J.S.; Birnbaum, Y.; Zanelli, C. & Luo, H. (2000). Noninvasive, Transthoracic, Low-frequency Ultrasound Augments Thrombolysis in a Canine Model of Acute Myocardial Infarction, *Circulation*, Vol.101, No.17 (May 2000) pp.2026-2029, ISSN 1524-4539

Siegel, J. & Luo, H. (2008). Ultrasound Thrombolysis, *Ultrasonics*, Vol.48, No.4, (August 2008) pp.312-320, ISSN 0041-624X

Siegel, R.J., Suchkova, V.N.; Miyamoto, T.; Luo, H.; Baggs, R.B.; Neuman, Y.; Horzewski, M.; Suorsa, V.; Kobal, S.; Thompson, T.; Echt D.; & Francis, C.W. (2004) Ultrasound Energy Improves Myocardial Perfusion in the Presence of Coronary Occlusion, *J. Am. Coll. Cardiol.*, Vol.44, No.7, (October 2004), pp.1454-1458, ISSN 0735-1097

Silva, C.W.D. (1999). *Vibration Fundamentals and Practice*, CRC Press, ISBN 0849318084, New York

Tschepe, J.; Aspidov, A.A.; Helfmann J. & Herrig, M. (1994). Acoustical Waves via Optical Fibers for Biomedical Applications, *Proc. SPIE- Biomedical Optoelectronic Devices and Systems*, Vol.2084, pp.133-143, ISBN 9780819413512, Budapest, Hungary, 01 September 01, 1993

Tsuda, Y.; Mori E. & Ueha, S. (1983). Experimental Study of Ultrasonic Surgical Knife, *Jpn. J. Appl. Phys.*, Vol.22, Supplement 22-3 (1983) pp.105-107, ISSN 0021-4922

Uchino, K. & Giniewicz J.R. (2003). *Micromechatronics*, CRC Press, ISBN 0824741099, New York

Wiksell, H.; Martin, H.; Coakham, H.; Berggren A. & Westermark, S. (2000). Miniaturized Ultrasonic Aspiration Handpiece for Increased Applicability, *Eur. J. Ultrasound*, Vol. 11, No.1, (March 2000), pp.41-46, ISSN 09298266

Woias, P.; Micropumps - Past, Progress and Future Prospects, *Sensors and Actuators B*, Vol.105, No.1, (February 2005), pp.28-38, ISSN 0925-4005

Yeo, C.Y.; Shim, W.K.; Wouterson, E.; Li, Tao & Ma, J. (2008). Piezoelectric Materials for Impedance Driven Micro-Channel Flow, *Functional Materials Letters*, Vo.1, No.3, (2008), pp.225-228

Young, F.R. (1999). *Cavitation*, Imperial College Press, ISBN 1860941982, Singapore

Zhang, A.L & Wei, Y.Q. (2010). Generation of Droplets for Lab-on-a-Piezoelectric-Substrate Utilizing Surface Acoustic Wave, *Proceedings of the 2010 symposium on piezoelectricity, acoustic waves, and device applications (SPAWDA 2010)*, pp.68-71, ISBN 978-1-4244-9822-2 ,Xiamen, China, December 10-13, 2010

Zhang, J.H.; Wang, D.K.; Wang, S.Y. & Qnuki, A. (2003). Research on Piezoelectric Pump-Lagging of Valve, *Chinese Journal of Mechanical Engineering*, Vol.39, No.5, (May 2003), pp.107-110, ISSN 05776686

2

Modeling and Investigation of One-Dimensional Flexural Vibrating Mechatronic Systems with Piezoelectric Transducers

Andrzej Buchacz and Marek Płaczek
Silesian University of Technology, Poland

1. Introduction

Piezoelectricity has found a lot of applications since it was discovered in 1880 by Pierre and Jacques Curie. There are many applications of the direct piezoelectric effect - the production of an electric potential when stress is applied to the piezoelectric material, as well as the reverse piezoelectric effect - the production of strain when an electric field is applied (Moheimani & Fleming, 2006). In this chapter analysis of mechatronic systems with both direct and reverse piezoelectric effects applications in mechatronic systems are presented. In considered systems piezoelectric transducers are used as actuators – the reverse piezoelectric effect application, or as vibration dampers with the external shunting electric circuit – the direct piezoelectric effect. In the first case piezoelectric transducers can be used as actuators glued on the surface of a mechanical subsystem in order to generate desired vibrations or also to control and damp vibrations in active damping applications (Kurnik et al., 1995; Gao & Liao, 2005). In this case electric voltage is generated by external control system and applied to the transducer. In the second case piezoelectric transducers are used as passive vibration dampers. A passive electric network is adjoined to transducer's clamps. The possibility of dissipating mechanical energy with piezoelectric transducers shunted with passive electric circuits was experimentally investigated and described in many publications (Buchacz & Płaczek, 2009a; Fein, 2008; Hagood & von Flotow, 1991; Kurnik, 2004). There are two basic applications of this idea. In the first method only a resistor is used as a shunting circuit and in the second method it is a passive electric circuit composed of a resistor and inductor. Many authors have worked to improve this idea. For example multimode piezoelectric shunt damping systems were described (Fleming et al., 2002). What is more there are many commercial applications of this idea (Yoshikawa et al., 1998).

Mechatronic systems with piezoelectric sensors or actuators are widely used because piezoelectric transducers can be applied in order to obtain required dynamic characteristic of designed system. It is very important to use very precise mathematical model and method of the system's analysis to design it correctly. It was proved that it is very important to take into consideration influence of all analyzed system's elements including a glue layer between piezoelectric transducer and mechanical subsystem (Pietrzakowski, 2001; Buchacz & Płaczek, 2010b). It is indispensable to take into account geometrical and material parameters of all system's components because the omission of the influence of one of them results in inaccuracy in the analysis of the system.

This work presents the issues of modeling and testing of flexural vibrating mechatronic systems with piezoelectric transducers used as actuators or vibration dampers. Analysis method of the considered system will be presented, started from development of the mathematical model, by setting its characteristics, to determine the influence of the system's properties on these characteristics.

The discussed subject is important due to increasing number of applications, both simple and reverse piezoelectric phenomena in various modern technical devices. The process of modeling of technical devices with piezoelectric materials is complex and requires large amounts of time because of the complexity of the phenomena occurring in these systems. The correct description of the system by its mathematical model during the design phase is fundamental condition for proper operation of designed system. Therefore, in the work the processes of modeling, testing and verification of used mathematical models of one-dimensional vibrating mechatronic systems will be presented. A series of discrete - continuous mathematical models with different simplifying assumptions will be created. Using created models and corrected approximate Galerkin method dynamic characteristics of considered systems will be designated. An analysis of influence of some geometrical and material parameters of system's components on obtained characteristics will be conducted. Mathematical model that provides the most accurate analysis of the system and maximum simplification of used mathematical tools and minimize required amount of time will be indicated. Identification of the optimal mathematical model that meets the assumed criteria is the main purpose of this work, which is an introduction to the task of synthesis of one-dimensional vibrating continuous systems.

2. Considered system with piezoelectric transducer and assumptions

The main aim of this work is to designate dynamic characteristics of a mechatronic system with piezoelectric transducer used as an actuator or passive vibration damper. It is a cantilever beam which has a rectangular constant cross-section, length l, width b and thickness h_b. Young's modulus of the beam is denoted E_b. A piezoelectric transducer of length l_p is bonded to the beam's surface within the distance of x_1 from a clamped end of the beam. The transducer is bonded by a glue layer of thickness h_k and Kirchhoff's modulus G. The glue layer has homogeneous properties in overall length. The system under consideration in both cases (with piezoelectric actuator or vibration damper) is presented in Fig. 1.

In order to analyze vibration of the systems following assumptions were made:

- material of which the system is made is subjected to Hooke's law,
- the system has a continuous, linear mass distribution,
- the system's vibration is harmonic,
- planes of cross-sections that are perpendicular to the axis of the beam remain flat during deformation of the beam - an analysis is based on the Bernoulli's hypothesis of flat cross-sections,
- displacements are small compared with the dimensions of the system.

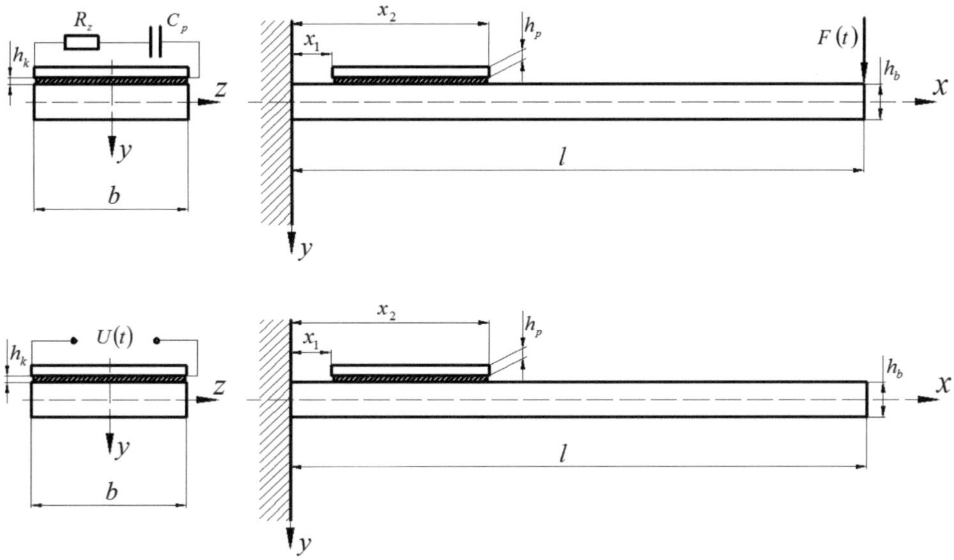

Fig. 1. Shape of the considered system: *a)* with piezoelectric passive vibration damper, *b)* with piezoelectric actuator

Structural damping of the beam and glue layer was taken into account in mathematical models of considered systems using Kelvin-Voigt model of material. It was introduced by replacing Young's modulus of the beam and modulus of elasticity in shear of the glue layer by equations:

$$E_b^* = E_b\left(1+\eta_b\frac{\partial}{\partial t}\right), \tag{1}$$

$$G^* = G\left(1+\eta_k\frac{\partial}{\partial t}\right), \tag{2}$$

where η_b and η_k denote structural damping coefficients of the beam and the glue layer that have time unit (Pietrzakowski, 2001).

It was assumed that the beam is made of steel and piezoelectric transducer is a PZT transducer. Geometric and material parameters of the system's elements: mechanical subsystem – the beam, the glue layer and the piezoelectric transducer are presented in tables 1, 2 and 3.

Geometric parameters	Material parameters
$l = 0{,}24[m]$	$E_b = 210000[MPa]$
$b = 0{,}04[m]$	$\rho_b = 7850\left[\frac{kg}{m^3}\right]$
$h_b = 0{,}002[m]$	$\eta_b = 8\cdot 10^{-5}[s]$

Table 1. Parameters of the mechanical subsystem

Geometric parameters	Material parameters
$x_1 = 0{,}01[m]$	$d_{31} = -240 \cdot 10^{-12}\left[\frac{m}{V}\right]$
$x_2 = 0{,}09[m]$	$e_{33}{}^{T} = 2900 \cdot \varepsilon_0\left[\frac{F}{m}\right]$
$h_p = 0{,}001[m]$	$s_{11}{}^{E} = \frac{1}{c_{11}{}^{E}} = 17 \cdot 10^{-12}\left[\frac{m^2}{N}\right]$
$b_p = 0{,}04[m]$	$\rho_p = 7450\left[\frac{kg}{m^3}\right]$

Table 2. Parameters of the piezoelectric transducer

Geometric parameters	Material parameters
$h_k = 0{,}0001[m]$	$G = 1000 \cdot 10^6 [Pa]$
	$\eta_k = 10^{-3}[s]$

Table 3. Parameters of the glue layer

Symbols ρ_b and ρ_p denote density of the beam and transducer. d_{31} is a piezoelectric constant, $e_{33}{}^{T}$ is a permittivity at zero or constant stress, $s_{11}{}^{E}$ is flexibility and $c_{11}{}^{E}$ is a Young's modulus at zero or constant electric field.

Dynamic characteristics of considered systems are described by equations:

$$y(x,t) = \alpha_Y \cdot F(t), \tag{3}$$

$$y(x,t) = \alpha_V \cdot U(t), \tag{4}$$

where $y(x,t)$ is the linear displacement of the beam's sections in the direction perpendicular to the beam's axis. In case of the system with piezoelectric vibration damper it is dynamic flexibility – relation between the external force applied to the system and beam's deflection (equation 3). In case of the system with piezoelectric actuator it is relation between electric voltage that supplies the actuator and beam's deflection (equation 4) (Buchacz & Płaczek, 2011). Externally applied force in the first system and electric voltage in the second system are described as:

$$F(t) = F_0 \cdot \cos\omega t, \tag{5}$$

$$U(t) = U_0 \cdot \cos\omega t, \tag{6}$$

and they were assumed as harmonic functions of time.

3. Approximate Galerkin method verification – Analysis of the mechanical subsystem

In order to designate dynamic characteristics of considered systems correctly it is important to use very precise mathematical model. Very precise method of the system's analysis is very important too. It is impossible to use exact Fourier method of separation of variables in analysis of mechatronic systems, this is why the approximate method must be used. To analyze considered systems approximate Galerkin method was chosen but verification of this method was the first step (Buchacz & Płaczek, 2010c). To check accuracy and verify if the Galerkin method can be used to analyze mechatronic systems the mechanical subsystem was analyzed twice. First, the exact method was used to designate dynamic flexibility of the mechanical subsystem. Then, the approximate method was used and obtained results were juxtaposed. The mechanical subsystem is presented in Fig.2.

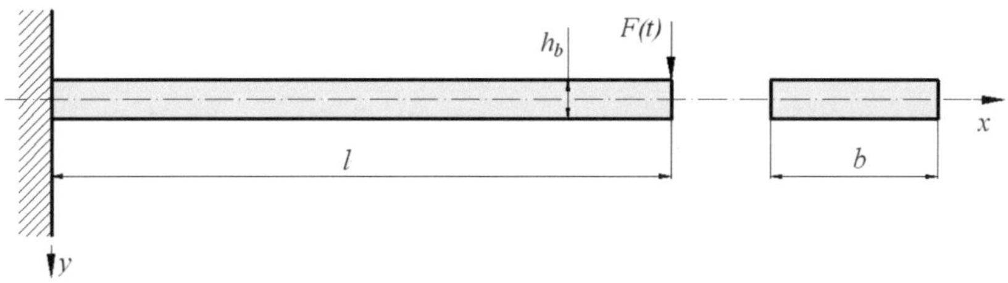

Fig. 2. Shape of the mechanical subsystem

The equation of free vibration of the mechanical subsystem was derived in agreement with d'Alembert's principle. The external force $F(t)$ was neglected. Taking into account equilibrium of forces and bending moments acting on the beam's element, after transformations a well known equation was obtained:

$$\frac{\partial^2 y(x,t)}{\partial t^2} = -a^4 \frac{\partial^4 y(x,t)}{\partial x^4}, \tag{7}$$

where:

$$a = \sqrt[4]{\frac{E_b J_b}{\rho_b A_b}}. \tag{8}$$

A_b and J_b are the area and moment of inertia of the beam's cross-section. In order to determine the solution of the differential equation of motion (7) Fourier method of separation of variables was used. Taking into account the system's boundary conditions, after transformations the characteristic equation of the mechanical subsystem was obtained:

$$\cos kl = -\frac{1}{\cosh kl}. \tag{9}$$

Graphic solution of the equation (9) is presented in Fig. 3. The solution of the system's characteristic equation approach to limit described by equation:

$$k_n = \frac{2n-1}{2l}\pi, \qquad n = 1,2,3... \tag{10}$$

This solution is precise for $n > 3$. For the lower values of n solutions should be readout from the graphic solution (Fig. 3) and they are presented in table 4.

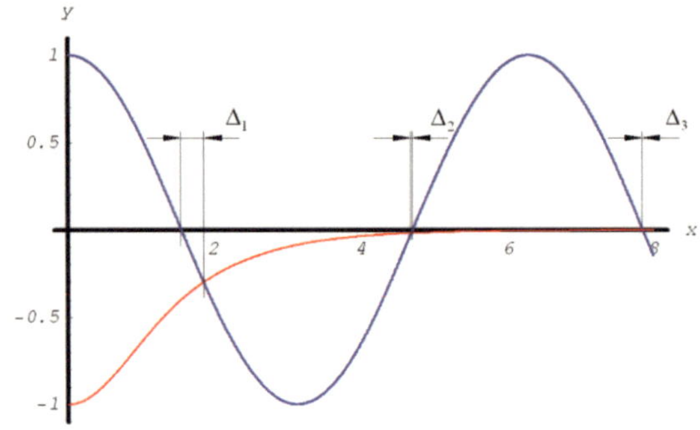

Fig. 3. The graphic solution of the characteristic equation of the system (equation 7)

Taking into account the system's boundary and initial conditions, after transformations the sequence of eigenfunctions is described by the equation:

$$X_n(x) = A_n \frac{\cos k_n l + \cosh k_n l}{\sin k_n l - \sinh k_n l}(\cos k_n x - \cosh k_n x) + \sin k_n x - \sinh k_n x. \tag{11}$$

Assuming zero initial conditions and taking into account that the deflection of the beam is a harmonic function with the same phase as the external force the final form of the solution of differential equation (7) can be described by the equation:

$$y_n(x,t) = \sum_{n=1}^{\infty} X_n(x) \cdot \cos \omega t, \tag{12}$$

and dynamic flexibility of the mechanical subsystem can be described as:

$$\alpha_Y = \frac{\left[\cosh \lambda^* l + \cos \lambda^* l\right]\left[-\sinh \lambda^* x + \sin \lambda^* x\right] + \left[\sinh \lambda^* l + \sin \lambda^* l\right]\left[\cosh \lambda^* x - \cos \lambda^* x\right]}{2 E_b J_b \lambda^{*3}\left[1 + \cos \lambda^* l \cosh \lambda^* l\right]}, \tag{13}$$

where:

$$\lambda^* = \sqrt[4]{\omega^2 \cdot \frac{\rho_b A_b}{E_b J_b}}. \tag{14}$$

In the approximate method the solution of differential equation (7) was assumed as a simple equation (Buchacz & Płaczek, 2009b, 2010d):

$$y(x,t) = A \sum_{n=1}^{\infty} \sin k_n x \cos \omega t, \tag{15}$$

where A is an amplitude of vibration. It fulfils only two boundary conditions - deflection of the clamped and free ends of the beam:

$$y(x,t) = 0\big|_{x=0}, \tag{16}$$

$$y(x,t) = A\big|_{x=l}. \tag{17}$$

The equation of the mechanical subsystem's vibration forced by external applied force can be described as:

$$\frac{\partial^2 y(x,t)}{\partial t^2} = -a^4 \frac{\partial^4 y(x,t)}{\partial x^4} + \frac{F(t)\delta(x-l)}{\rho_b A_b}. \tag{18}$$

Distribution of the external force was determined using Dirac delta function $\delta(x\text{-}l)$.

Corresponding derivatives of the assumed approximate solution of the differential equation of motion (15) were substituted in the equation of forced beam's vibration (18). Taking into account the definition of the dynamic flexibility (3), after transformations absolute value of the dynamic flexibility of the mechanical subsystem (denoted Y) was determined:

$$Y = \left| \sum_{n=1}^{\infty} \frac{\delta(x-l)}{\rho_b A_b \left(-\omega^2 + a^4 {k_n}^4\right)} \right|. \tag{19}$$

Taking into account geometrical and material parameters of the considered mechanical subsystem (see table 1), the dynamic flexibility for the first three natural frequencies are presented in Fig. 4. In this figure results obtained using the exact and the approximate

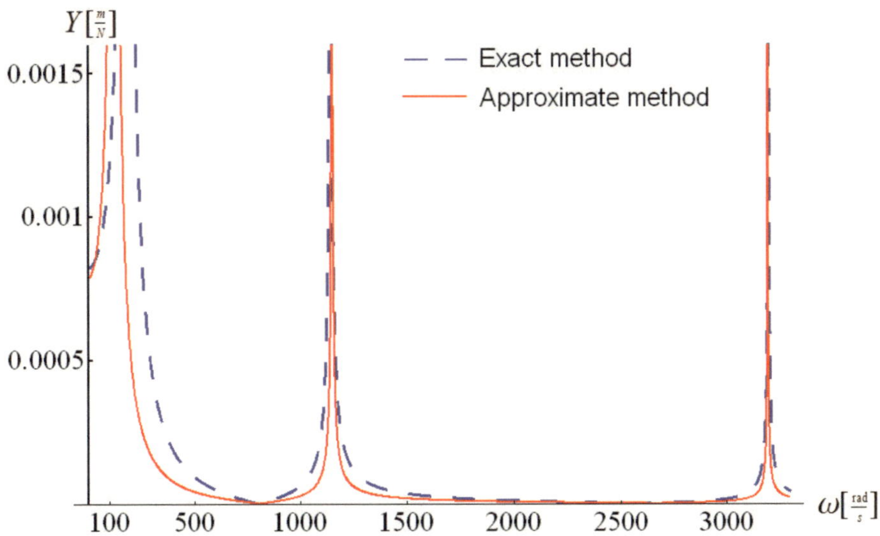

Fig. 4. The dynamic flexibility of the mechanical subsystem - exact and approximate method, for n=1,2,3

n	The exact method	The approximate method	$\Delta[\%]$
1	$k_1 = \frac{1,8751}{l}$	$k_1 = \frac{\pi}{2l}$	29,8
2	$k_2 = \frac{4,6941}{l}$	$k_2 = \frac{3\pi}{2l}$	-0,782
3	$k_3 = \frac{7,85477}{l}$	$k_3 = \frac{5\pi}{2l}$	0,023
>3	$k_n = (2n-1)\frac{\pi}{2l}$		0

Table 4. The first three roots of the characteristic equation and shifts of values of the system's natural frequencies

methods are juxtaposed. Inexactness of the approximate method is very meaningful for the first three natural frequencies. Shifts of values of the system's natural frequencies are results of the discrepancy between the assumed solution of the system's differential equation of motion in the approximate method and solution obtained on the basis of graphic solution of the system's characteristic equation in the exact method. These discrepancies are shown in table 4. So it is possible to identify discrepancies between the exact and approximate methods without knowing any geometrical and material parameters. Knowing the characteristic equation of the mechanical system with known boundary conditions and assumed solution of the differential equation of motion it is possible to determine whether the solution obtained using the approximate method differs from the exact solution.

The approximate method was corrected for the first three natural frequencies of the considered system by introduction in equation (19) correction coefficients described by the equation:

$$\Delta\omega_n = \omega_n - \omega_n', \tag{20}$$

where ω_n and ω_n' are values obtained using the exact and approximate methods, respectively (Buchacz & Płaczek, 2010c). The dynamic flexibility of the mechanical subsystem before and after correction is presented in Fig. 5 separately for the first three natural frequencies.

Results of assumption of simplified eigenfunction of variable x (equation 15) are also inaccuracies of the system's vibration forms presented in Fig. 6.

The approximate Galerkin method with corrected coefficients gives a very high accuracy and obtained results can be treated as very precise (see Fig. 5). So it can be used to analyze mechatronic systems with piezoelectric transducers. The considered system – a cantilever beam was chosen purposely because inexactness of the approximate Galerkin method is the biggest in this way of the system fixing.

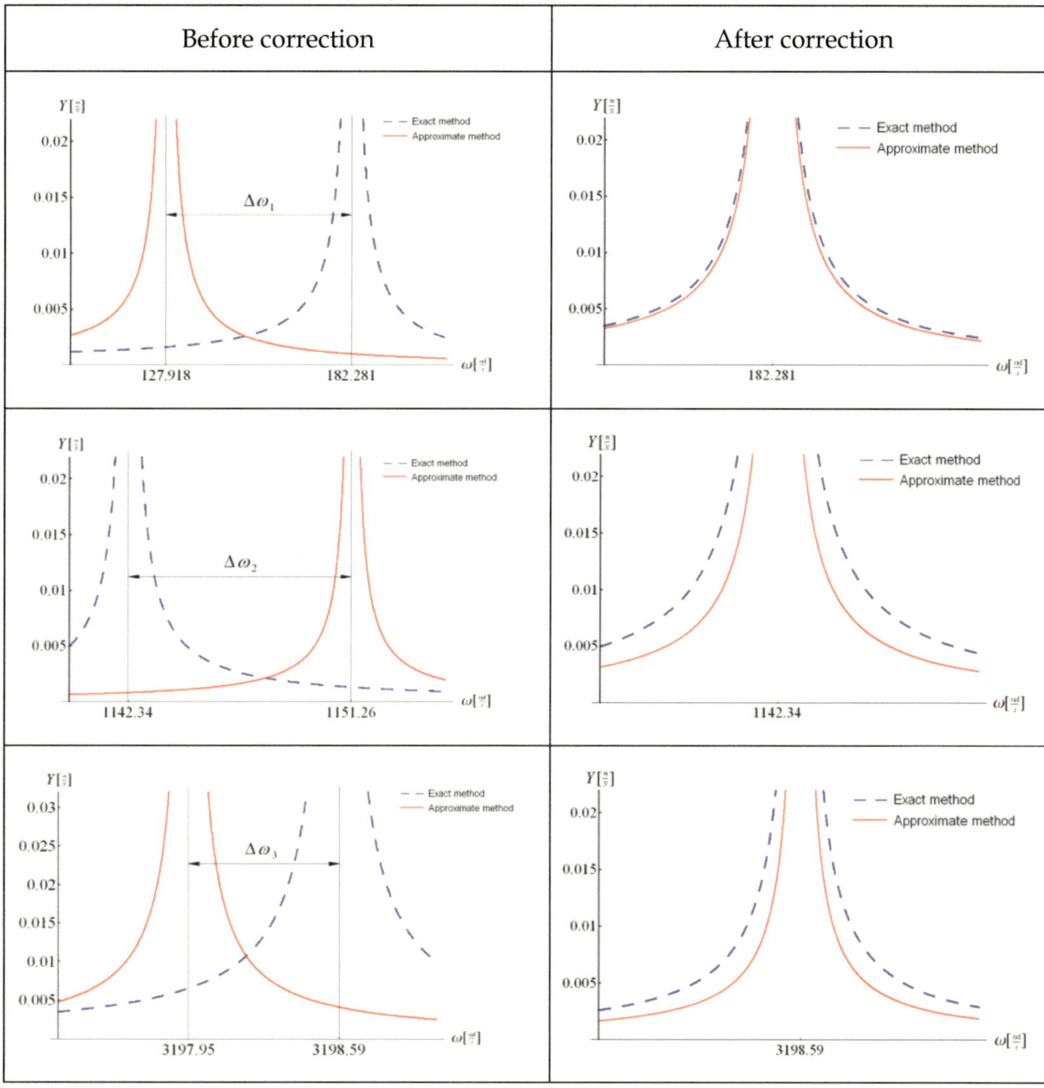

Fig. 5. The dynamic flexibility of the mechanical subsystem - exact and approximate method before and after correction.

4. Mechatronic system with broad-band piezoelectric vibration damper

The considered mechatronic system with broad-band, passive piezoelectric vibration damper was presented in Fig. 1. In this case, to the clamps of a piezoelectric transducer, an external shunt resistor with a resistance R_Z is attached. As a result of the impact of vibrating beam on the transducer and its strain the electric charge and additional stiffness of electromechanical nature, that depends on the capacitance of the piezoelectric transducer, are generated. Electricity is converted into heat and give up to the environment. Piezoelectric transducer with an external resistor is called a shunt broad-band damper (Buchacz & Płaczek, 2010c; Hagood & von Flotow, 1991;Kurnik, 1995).

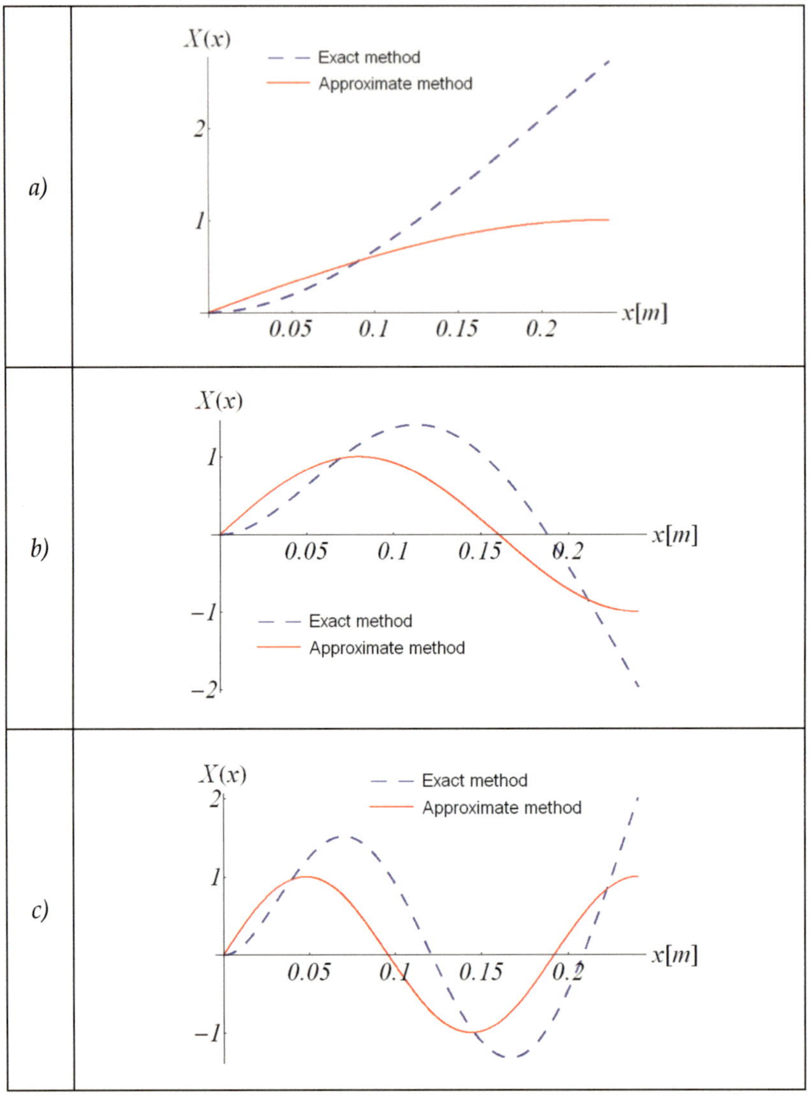

Fig. 6. Vibration forms of the mechanical subsystem – the exact and approximate methods, a) the first natural frequency, b) the second natural frequency, c) the third natural frequency

Piezoelectric transducer can be described as a serial connection of a capacitor with capacitance C_P, internal resistance of the transducer R_P and strain-dependent voltage source U_P. However, it is permissible to assume a simplified model of the transducer where internal resistance is omitted. In this case internal resistance of the transducer, which usually is in the range 50 – 100 Ω (Behrens & Fleming, 2003) is negligibly small in comparison to the resistance of externally applied electric circuit (400 kΩ), so it was omitted. Taking into account an equivalent circuit of the piezoelectric transducer presented in Fig. 7, an electromotive force generated by the transducer and its electrical capacity are treated as a serial circuit. The considered mechatronic system can be represented in the form, as shown in Fig. 1. So, the piezoelectric transducer with an external shunt resistor is treated as a serial

RC circuit with a harmonic voltage source generated by the transducer (Behrens & Fleming, 2003; Moheimani & Fleming, 2006).

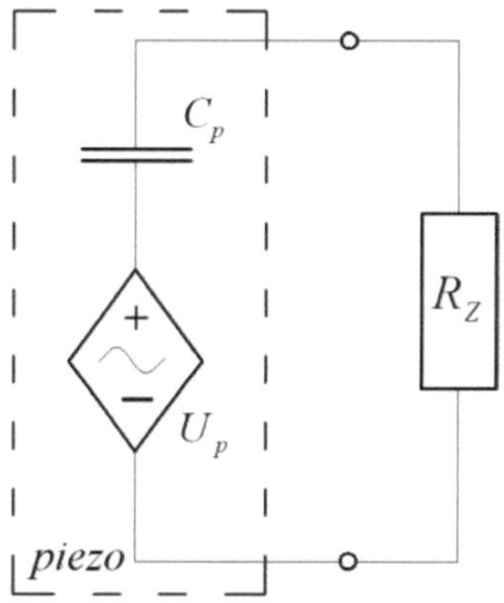

Fig. 7. The substitute scheme of the piezoelectric transducer with an external shunt resistor

4.1 A series of mathematical models of the mechatronic system with piezoelectric vibration damper

A series of mathematical models of the considered mechatronic system with broad-band, passive piezoelectric vibration damper was developed. Different type of the assumptions and simplifications were introduced so developed mathematical models have different degree of precision of real system representation. A series of discrete - continuous mathematical models was created. The aim of this study was to develop mathematical models of the system under consideration, their verification and indication of adequate model to accurately describe the phenomena occurring in the system and maximally simplify the mathematical calculations and minimize required time (Buchacz & Płaczek, 2009b, 2010b).

4.1.1 Discrete – continuous mathematical model with an assumption of perfectly bonded piezoelectric damper

In the first mathematical model of the considered mechatronic system there is an assumption of perfectly bonded piezoelectric transducer - strain of the transducer is exactly the same as the beam's surface strain. Taking into account arrangement of forces and bending moments acting in the system that are presented in Fig. 8, differential equation of motion can be described as:

$$\frac{\partial^2 y(x,t)}{\partial t^2} = -a^4\left(1+\eta_b\frac{\partial}{\partial t}\right)\frac{\partial^4 y(x,t)}{\partial x^4} + \frac{1}{\rho_b A_b}\frac{\partial^2 M_p(x,t)}{\partial x^2} + \frac{\delta(x-l)}{\rho_b A_b}F(t). \tag{21}$$

$T(x,t)$ denotes transverse force, $M(x,t)$ bending moment and $M_P(x,t)$ bending moment generated by the piezoelectric transducer that can be described as:

$$M_p(x,t) = \frac{h_b + h_p}{2} \cdot F_p(t). \tag{22}$$

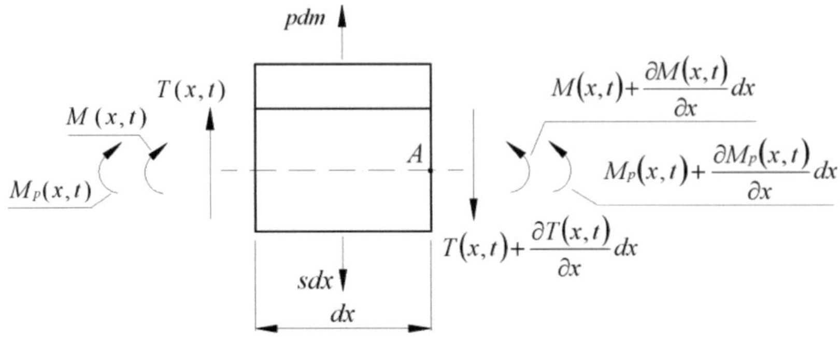

Fig. 8. Arrangement of forces and bending moments acting on the cut out part of the beam and the piezoelectric transducer with length *dx*

Piezoelectric materials can be described by a pair of constitutive equations witch includes the relationship between mechanical and electrical properties of transducers (Preumont, 2006; Moheimani & Fleming, 2006). In case of the system under consideration these equations can be written as:

$$D_3 = \varepsilon_{33}{}^T E_3 + d_{31} T_1, \tag{23}$$

$$S_1 = d_{31} E_3 + s_{11}{}^E T_1. \tag{24}$$

Symbols $\varepsilon_{33}{}^T$, d_{31}, $s_{11}{}^E$ are dielectric, piezoelectric and elasticity constants. Superscripts T and E denote value at zero/constant stress and zero/constant electric field, respectively. Symbols D_3, S_1, T_1 and E_3 denote electric displacement, strain, stress and the electric field in the directions of the axis described by the subscript. After transformation of equation (24), force generated by the transducer can be described as:

$$F_P(t) = c_{11}{}^E A_p \left[S_1(x,t) - \lambda_1(t) \right], \tag{25}$$

where:

$$\lambda_1(t) = d_{31} \cdot E_3 = d_{31} \frac{U_C(t)}{h_p}. \tag{26}$$

Symbol $c_{11}{}^E$ denotes Young's modulus of the transducer at zero/constant electric field (inverse of elasticity constant). $U_C(t)$ is an electric voltage on the capacitance C_p. Due to the fact that the piezoelectric transducer is attached to the surface of the beam on the section from x_1 to x_2 its impact was limited by introducing Heaviside function $H(x)$. Finally, equation (21) can be described as:

$$\frac{\partial^2 y(x,t)}{\partial t^2} = -a^4 \left(1 + \eta_b \frac{\partial}{\partial t} \right) \frac{\partial^4 y(x,t)}{\partial x^4} + c_1 \cdot \frac{\partial^2}{\partial x^2} \left[H \cdot S_1(x,t) - H \cdot \lambda_1(t) \right] + \alpha F(t), \tag{27}$$

where:

$$c_1 = \frac{\left(h_b + h_p \right) \cdot c_{11}{}^E A_p}{2 \rho_b A_b}, \tag{28}$$

$$\alpha = \frac{\delta(x-l)}{\rho_b A_b}, \tag{29}$$

$$H = H(x-x_1) - H(x-x_2). \tag{30}$$

Equation of the piezoelectric transducer with external electric circuit can be described as:

$$R_Z C_P \frac{\partial U_C(t)}{\partial t} + U_C(t) = U_p(t), \tag{31}$$

where: C_P is the transducer's capacitance, $U_P(t)$ denotes electric voltage generated by the transducer as a result of its strain. Voltage generated by the transducer is a quotient of generated electric charge and capacitance of the transducer. After transformation of the constitutive equations (23) and (24) electric charge generated by the transducer can be described as (Kurnik, 2004):

$$Q(t) = \frac{l_p b d_{31}}{s_{11}^{E}} S_1(x,t) + l_p b \varepsilon_{33}^{T} \frac{U_C(t)}{h_p}\left(1-k_{31}^{2}\right), \tag{32}$$

where:

$$k_{31}^{2} = \frac{d_{31}^{2}}{s_{11}^{E} \varepsilon_{33}^{T}}, \tag{33}$$

is an electromechanical coupling constant that determines the efficiency of conversion of mechanical energy into electrical energy and electrical energy into mechanical energy of the transducer, whose value usually is from 0,3 to 0,7 (Preumont, 2006). Equation (33) describes the electric charge accumulated on the surface of electrodes of the transducer with an assumption about uniaxial, homogeneous strain of the transducer. Assuming an ideal attachment of the transducer to the beam's surface its strain is equal to the beam's surface strain and can be described as:

$$S_1(x,t) = \frac{h_b}{2} \cdot \frac{\partial^2 y(x,t)}{\partial x^2}. \tag{34}$$

Finally, equation (31) can be described as:

$$R_Z C_P \frac{\partial U_C(t)}{\partial t} + U_C(t) = \frac{l_p b d_{31}}{C_p s_{11}^{E}} S_1(x,t) + l_p b \varepsilon_{33}^{T} \frac{U_C(t)}{C_p h_p}\left(1-k_{31}^{2}\right). \tag{35}$$

Using the classical method of analysis of linear electric circuits and due to the low impact of the transient component on the course of electric voltage generated on the capacitance of the linear RC circuit the electric voltage $U_C(t)$ was assumed as:

$$U_C(t) = \frac{|U_p|}{\omega C_P |Z|} \cdot \sin(\omega t + \varphi), \tag{36}$$

where $|Z|$ and φ are absolute value and argument of the serial circuit impedance.

Equations (27) and (35) form a discrete-continuous mathematical model of the considered system.

4.1.2 Discrete – continuous mathematical model with an assumption of pure shear of a glue layer between the piezoelectric damper and beam's surface

Concerning the impact of the glue layer between the transducer and the beam's surface, the mathematical model of the system under consideration was developed. It will allow more detailed representation of the real system. First, a pure shear of the glue layer was assumed. Arrangement of forces and bending moments acting in the system modeled with this assumption is presented in Fig. 9.

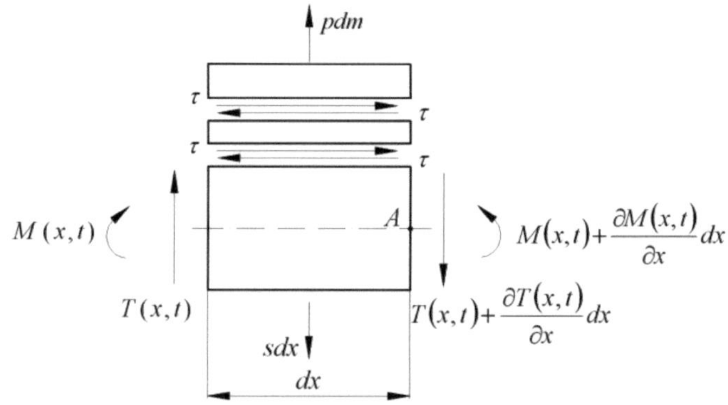

Fig. 9. Forces and bending moments in case of the pure shear of the glue layer

Shear stress was determined according to the Hook's law, assuming small values of pure non-dilatational strain:

$$\tau = \frac{\Delta l}{h_k} \cdot G. \tag{37}$$

Δl is a displacement of the lower and upper surfaces of the glue layer. Movements of the beam, the glue layer and the transducer are shown in Fig. 10.

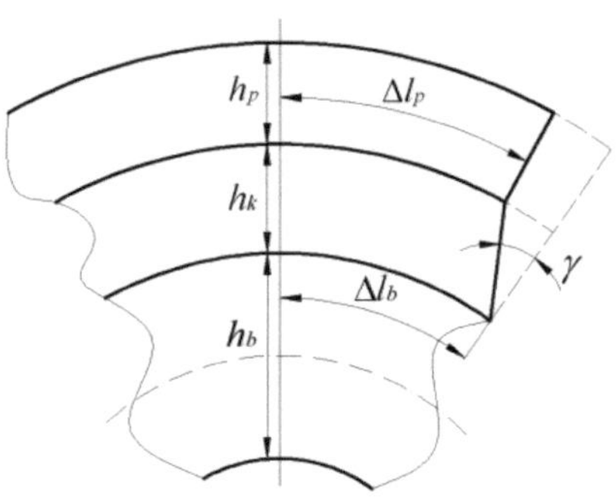

Fig. 10. Movements of the beam, the glue layer and the piezoelectric transducer in the case of pure shear of the glue layer

Uniform distribution of shear stress along the glue layer was assumed. The real strain of the transducer is a difference of the glue layer's upper surface strain and the free transducer's strain that is a result of electric field on the transducer's electrodes, so Δl can be described as:

$$\Delta l = l_p\left[\varepsilon_b(x,t) - \varepsilon_k(x,t) + \lambda_1(t)\right], \tag{38}$$

where: ε_b and ε_k are the beam's and the glue layer's upper surfaces strains.

Finally, obtained system of equations:

$$\begin{cases} \dfrac{\partial^2 y(x,t)}{\partial t^2} = -a^4\left(1+\eta_b\dfrac{\partial}{\partial t}\right)\dfrac{\partial^4 y(x,t)}{\partial x^4} + c_2\left(1+\eta_k\dfrac{\partial}{\partial t}\right)\dfrac{\partial}{\partial x}H\left[\varepsilon_b(x,t)-\varepsilon_k(x,t)+\lambda_1(t)\right]+\alpha F(t) \\ R_Z C_P\dfrac{\partial U_C(t)}{\partial t} + U_C(t) = \dfrac{l_p b d_{31}}{C_p s_{11}^{E}}S_1(x,t) + \dfrac{l_p b \varepsilon_{33}^{T}}{C_p h_p}\left(1-k_{31}^2\right)\cdot U_C(t) \end{cases}, \tag{39}$$

where:

$$c_2 = \frac{G l_p}{2\rho_b h_k}. \tag{40}$$

is the discrete-continuous mathematical model of the system under consideration with the assumption about pure shear of the glue layer.

4.1.3 Discrete – continuous mathematical model taking into account a shear stress and eccentric tension of a glue layer between the piezoelectric damper and beam's surface

In the next mathematical model the system under consideration was modeled as a combined beam in order to unify parameters of all components (Buchacz & Płaczek, 2009c). Shear stress and eccentric tension of the glue layer were assumed. The substitute cross-section of considered system presented in Fig. 11 was introduced by multiplying the width of the piezoelectric transducer and the glue layer by factors:

$$m_p = \frac{c_{11}^{E}}{E_b}, \tag{41}$$

$$m_k = \frac{2G(1+\nu)}{E_b}. \tag{42}$$

Symbol υ denotes the Poisson's ratio of the glue layer.

Taking into account the eccentric tension of the glue layer under the action of forces presented in Fig.12 the stress on the substitute cross-section's surfaces was assigned.

$F_P(t)$ denotes force generated by the piezoelectric transducer and $F_b(t)$ denotes forces generated by the bending beam as a result of the beam's elasticity. The area, location of the central axis and moment of inertia of the substitute cross-section were calculated:

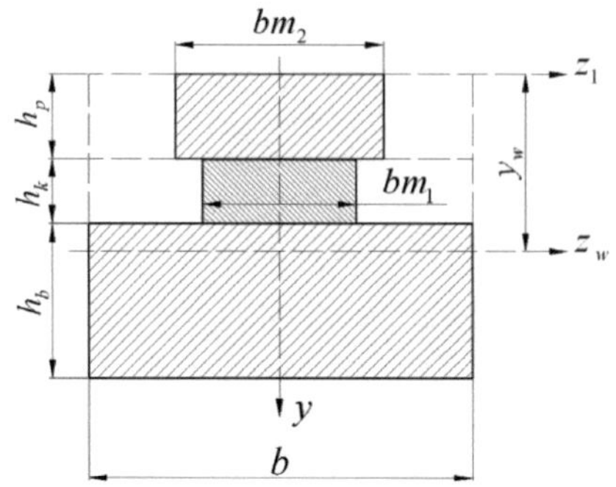

Fig. 11. Position of the center of gravity of substitute cross-section of the beam

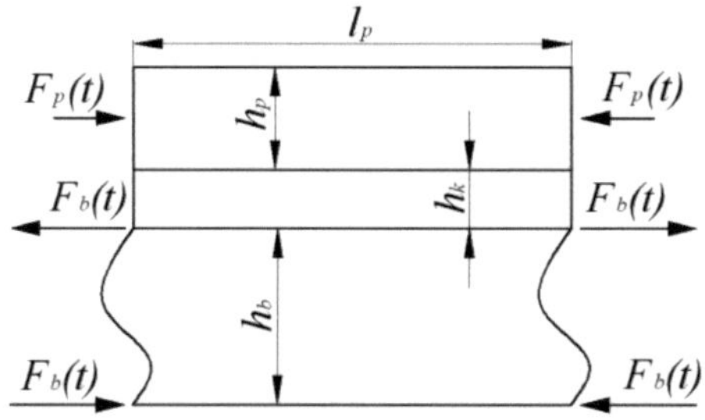

Fig. 12. Arrangement of forces in case of eccentric tension of the glue layer

$$A_w = bh_b + m_k bh_k + m_p bh_p, \tag{43}$$

$$y_w = A_w^{-1} b\left[h_b\left(h_k + h_p + \frac{h_b}{2}\right) + m_k h_k\left(h_p + \frac{h_k}{2}\right) + m_p \frac{h_p^2}{2}\right], \tag{44}$$

$$J_w = b\left\{h_b\left[\frac{h_b^2}{12} + \left(h_p + h_k + \frac{h_b}{2} - y_w\right)^2\right] + m_k h_k\left[\frac{h_k^2}{12} + \left(y_w - h_p - \frac{h_k}{2}\right)^2\right] + m_p h_p\left[\frac{h_p^2}{12} + \left(y_w - \frac{h_p}{2}\right)^2\right]\right\}, \tag{45}$$

were calculated and stress on the surfaces of the substitute cross-section was assigned:

$$\sigma_i = m_i\left[F_b(t)\left(-\frac{h_b y}{J_w}\right) - F_p(t)\left(\frac{1}{A_w} - \frac{\left(-y_w + 0{,}5h_p\right)y}{J_w}\right)\right], \tag{46}$$

where subscript i denotes element of the composite beam (i=b,k,p). In case of the beam, value of the symbol m_b is equal to one, while in case of the transducer and the glue layer m_p and m_k are described by equations (41) and (42). Using the basic laws and dependences from theory of strength of materials the real strain of the piezoelectric transducer was assigned:

$$S_1(x,t) = W_1 \cdot \varepsilon_b(x,t) - W_2 \cdot \lambda_1(t), \tag{47}$$

where:

$$W_1 = \frac{h_p - y_w}{\left(h_p + h_k - y_w\right)\left[1 - \frac{E_p A_p}{E_b A_w}\left(\frac{h_p - y_w}{h_p + h_k - y_w} - 1\right)\right]}, \tag{48}$$

$$W_2 = \frac{E_p \frac{A_p}{A_w}\left(\frac{h_p - y_w}{h_p + h_k - y_w} - 1\right)}{E_b\left[1 - \frac{E_p A_p}{E_b A_w}\left(\frac{h_p - y_w}{h_p + h_k - y_w} - 1\right)\right]}. \tag{49}$$

To determine the value of shear stress on the plane of contact of the transducer and beam the following dependence was used:

$$\tau(x,y) = \frac{T(x,t)\cdot S_z(y)}{J_w \cdot b(y)}, \tag{50}$$

where: $S_Z(y)$ is a static moment of cut off part of the composite beam's cross-section relative to the weighted neutral axis. Transverse force $T(x,t)$ can be calculated as a derivative of bending moment acting on the beam's cross-section:

$$T(x,t) = \frac{\partial\left\{H\left[W_3 \cdot \varepsilon_b(x,t) - W_4 \cdot \lambda_1(t)\right]\right\}}{\partial x}, \tag{51}$$

where:

$$W_3 = c_{11}{}^E A_p W_1\left(y_w - \frac{h_p}{2}\right) + \frac{h_b E_b J_w}{h_b\left(y_w - h_p - h_k\right)} + \frac{h_b W_1 c_{11}{}^E A_p J_w}{h_b\left(y_w - h_p - h_k\right)}\left[A_w{}^{-1} + \frac{0{,}5h_p - y_w}{J_w}\left(-y_w + h_p + h_k\right)\right], \tag{52}$$

$$W_4 = c_{11}{}^E A_p\left(W_2 + 1\right)\cdot\left(y_w - \frac{h_p}{2}\right) + \frac{\left(W_2 + 1\right)c_{11}{}^E h_b A_p J_w}{h_b\left(y_w - h_p - h_k\right)}\left[A_w{}^{-1} + \frac{0{,}5h_p - y_w}{J_w}\left(-y_w + h_p + h_k\right)\right]. \tag{53}$$

Finally, the discrete-continuous mathematical model of the system can be described as:

$$\begin{cases} \frac{\partial^2 y(x,t)}{\partial t^2} = -a^4\left(1 + \eta_b \frac{\partial}{\partial t}\right)\frac{\partial^4 y(x,t)}{\partial x^4} + \frac{S_z(y)}{2\rho_b J_w b}\left(1 + \eta_k \frac{\partial}{\partial t}\right)\frac{\partial^2}{\partial x^2}\left[HW_3\varepsilon_b(x,t) - HW_4\lambda_1(t)\right] + \alpha F(t) \\ R_Z C_P \frac{\partial U_C(t)}{\partial t} + U_C(t) = \frac{l_p b d_{31}}{C_p s_{11}{}^E} S_1(x,t) + \frac{l_p b \varepsilon_{33}{}^T}{C_p h_p}\left(1 - k_{31}{}^2\right)\cdot U_C(t) \end{cases}. \tag{54}$$

Obtained system of equations is a mathematical model with assumptions of shear stress and eccentric tension of the glue layer.

4.1.4 Discrete – continuous mathematical model taking into account a bending moment generated by the transducer and eccentric tension of a glue layer between the piezoelectric damper and beam's surface

Taking into account parameters of the combined beam introduced in section 4.1.4 the discrete-continuous mathematical model with influence of the glue layer on the dynamic characteristic of the system was developed. However, in this model the impact of the piezoelectric transducer was described as a bending moment, similarly as in the mathematical model with the assumption of perfectly attachment of the transducer. Homogeneous, uniaxial tension of the transducer was assumed and its deformation was described by the equation (47). In this case the bending moment generated by the transducer can be described as:

$$M_p(x,t)=\left(\frac{h_p+h_b}{2}+h_k\right)c_{11}{}^E A_p\left[W_1\varepsilon_b(x,t)-(W_2+1)\lambda_1(t)\right]. \tag{55}$$

Obtained system of equations:

$$\begin{cases} \dfrac{\partial^2 y(x,t)}{\partial t^2}=-a^4\left(1+\eta_b\dfrac{\partial}{\partial t}\right)\dfrac{\partial^4 y(x,t)}{\partial x^4}+c_4\dfrac{\partial^2}{\partial x^2}\left[T_1\varepsilon_b(x,t)H-(T_2+1)\lambda_1(t)H\right]+\alpha F(t) \\ R_Z C_P\dfrac{\partial U_C(t)}{\partial t}+U_C(t)=\dfrac{l_p b d_{31}}{C_p s_{11}{}^E}S_1(x,t)+\dfrac{l_p b\varepsilon_{33}{}^T}{C_p h_p}\left(1-k_{31}{}^2\right)\cdot U_C(t) \end{cases}, \tag{56}$$

where:

$$c_4=\left(\frac{h_p+h_b}{2}+h_k\right)\frac{c_{11}{}^E A_p}{\rho_b A_b}, \tag{57}$$

is the discrete-continuous mathematical model of the system under consideration.

4.2 Dynamic flexibility of the system with broad-band piezoelectric vibration damper

Dynamic flexibility of the considered system was assigned using corrected approximate Galerkin method. Solution of the differential equation of the beam's motion with piezoelectric damper was assumed as a product of the system's eigenfunctions in accordance with the equation (15). For all mathematical models analogous calculations were done, therefore, an algorithm used to determine the dynamic flexibility of the system using the first mathematical model is presented. Obtained results for all mathematical models are presented in graphical form.

In the mathematical model of the considered mechatronic system with the assumption of perfectly bonded piezoelectric damper - equations (27) and (35) the derivatives of the approximate equation (15) were substituted. Assuming that the dynamic flexibility will be assigned on the free end of the beam ($x=l$), after transformations and simplifications a system of equations was obtained:

$$\begin{cases} P_1 A\cos\omega t + P_2 A\sin\omega t = P_3 B\sin\omega t + P_4 B\cos\omega t + \alpha F_0\cos\omega t \\ -P_5 A\cos\omega t + P_6 B\sin\omega t + P_7 B\cos\omega t = 0 \end{cases}, \tag{58}$$

where:

$$B = \frac{|U_p|}{\omega C_P |Z|}, \tag{59}$$

$$P_1 = \sin k_n l\left(-\omega^2 + a^4 k_n{}^4 - c_1 r_1 k_n{}^2 H''\right), \tag{60}$$

$$P_2 = -a^4 k_n{}^4 \eta_b \omega \sin k_n l, \tag{61}$$

$$P_3 = -\frac{c_1 d_{31} H''}{h_p}\cos\varphi, \tag{62}$$

$$P_4 = -\frac{c_1 d_{31} H''}{h_p}\sin\varphi, \tag{63}$$

$$P_5 = \frac{l_p b d_{31} r_1 k_n{}^2 \sin k_n l}{s_{11}{}^E R_Z C_P{}^2}, \tag{64}$$

$$P_6 = \frac{1}{R_Z C_P}\cos\varphi\left(1 - \frac{Z}{C_P}\right) - \omega\sin\varphi, \tag{65}$$

$$P_7 = \frac{1}{R_Z C_P}\sin\varphi\left(1 - \frac{Z}{C_P}\right) + \omega\cos\varphi. \tag{66}$$

Using mathematical dependences:

$$e^{\pm i\omega t} = \cos\omega t \pm i\cdot\sin\omega t, \tag{67}$$

$$\sin\omega t = \cos\left(\omega t - \frac{\pi}{2}\right), \tag{68}$$

after transformations the system of equations (58) can be written in matrix form:

$$\begin{bmatrix} P_1 - iP_2 & iP_3 - P_4 \\ -P_5 & -iP_6 + P_7 \end{bmatrix} \cdot \begin{bmatrix} A \\ B \end{bmatrix} = \begin{bmatrix} \alpha F_0 \\ 0 \end{bmatrix}. \tag{69}$$

Using Cramer's rule amplitude of the system's vibration can be calculated as:

$$A = \frac{W_A}{W}, \tag{70}$$

where W is a main matrix determinant and W_A is a determinant of the matrix formed by replacing the first column in the main matrix by the column vector of free terms. Obtained equation can be substituted in the assumed solution of the derivative equation of the beam's motion (15). Finally, in agreement with definition (3), the dynamic flexibility of the system under consideration can be described as:

$$\alpha_Y = \sum_{n=1}^{\infty} \frac{(\alpha P_7 - i\alpha P_6)\sin k_n l}{P_1 P_7 - P_2 P_6 - P_4 P_5 + i(P_3 P_5 - P_2 P_7 - P_1 P_6)}. \tag{71}$$

In order to eliminate complex numbers in equation (71) its numerator and denominator were multiplied by the number conjugate with the denominator. Absolute value of the obtained complex number was calculated:

$$Y = \sum_{n=1}^{\infty} \frac{\sqrt{R_1^2 + R_2^2}}{(P_1 P_7 - P_2 P_6 - P_4 P_5)^2 + (P_3 P_5 - P_2 P_7 - P_1 P_6)^2}, \tag{72}$$

where:

$$R_1 = \alpha \sin k_n l \left(P_1 P_7^2 - P_4 P_5 P_7 - P_3 P_5 P_6 + P_1 P_6^2 \right), \tag{73}$$

$$R_2 = \alpha \sin k_n l \left(P_2 P_7^2 - P_3 P_5 P_7 + P_4 P_5 P_6 + P_2 P_6^2 \right). \tag{74}$$

Taking into account geometrical and material parameters of the considered system presented in tables 1, 2 and 3, graphical solution of the equation (72) is presented in Fig. 13.

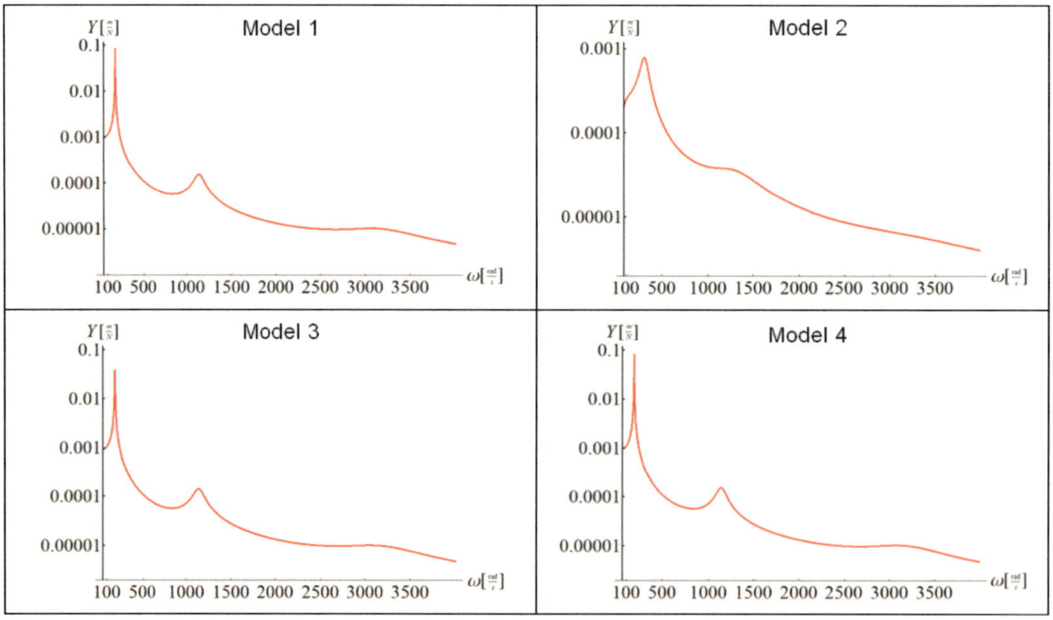

Fig. 13. Absolute value of the dynamic flexibility of mechatronic system with piezoelectric vibration damper, for the first three natural frequencies (in a half logarithmic scale)

Results obtained using the others mathematical models of the considered system are also presented in Fig. 13.

Using developed mathematical models and corrected approximate Galerkin method, very similar course of dynamic characteristics were obtained, except the second mathematical model with the assumption about pure shear of the glue layer. Shift of the natural frequencies in the direction of higher values of the mechatronic system in the direction of higher values can be observed. This shift is a result of increased stiffness of mechatronic system compared with the mechanical subsystem.

5. Mechatronic system with piezoelectric actuator

Developed mathematical models of the system with piezoelectric vibration damper were used to analyze the mechatronic system with piezoelectric actuator. In this case inverse piezoelectric effect is applied. Strain of the piezoelectric transducer is a result of externally applied electric voltage described by the equation (6). The considered system is presented in Fig. 1. Its parameters are presented in tables 1, 2 and 3. The aim of the system's analysis is to designate dynamic characteristic that is a relation between parameters of externally applied voltage and deflection of the free end of the beam (it was assumed that $x=l$), described by the equation (4).

In this case the internal capacitance C_P and resistance R_P of the piezoelectric transducer were taken into account, so transducer supplied by the external harmonic voltage source can be treated as a serial RC circuit with harmonic voltage source and was described by the equation (Buchacz & Płaczek, 2011):

$$R_P C_P \frac{\partial U_C(t)}{\partial t} + U_C(t) = U(t). \tag{75}$$

Equations of motion of the beam with piezoelectric actuator for all developed mathematical models were designated in agreement with d'Alembert's principle similarly as in the case of mechatronic system with piezoelectric vibration damper. Obtained absolute value of dynamic characteristics for all mathematical models of the system are presented in Fig. 14.

Final results are very similar for all mathematical models, except the second model with the assumptions about pure shear of the glue layer, as it was in case of analysis of system with piezoelectric vibration damper.

6. Analysis of influence of parameters of considered systems on dynamic characteristics

Developed mathematical models of considered systems were used to analyze influence of geometric and material parameters of systems on obtained dynamic characteristics. This study was carried out in dimensionless form in order to generalize obtained results. Results are presented in the form of three-dimensional graphs that show the course of the dimensionless absolute value of dynamic characteristic in relation to dimensionless frequency of externally applied force or electric voltage and one of the system's parameters dimensionless value. Dimensionless values of dynamic characteristics were introduced as:

$$Y_W = YE_b\sqrt{A_b}\left[\frac{s^2}{kg}\cdot m\cdot\frac{kg}{m\cdot s^2}=1\right], \tag{76}$$

$$Y_{VW} = \frac{Y_V}{\sqrt{d_{31}}}\left[\frac{m}{V}\cdot\frac{V}{m}=1\right]. \tag{77}$$

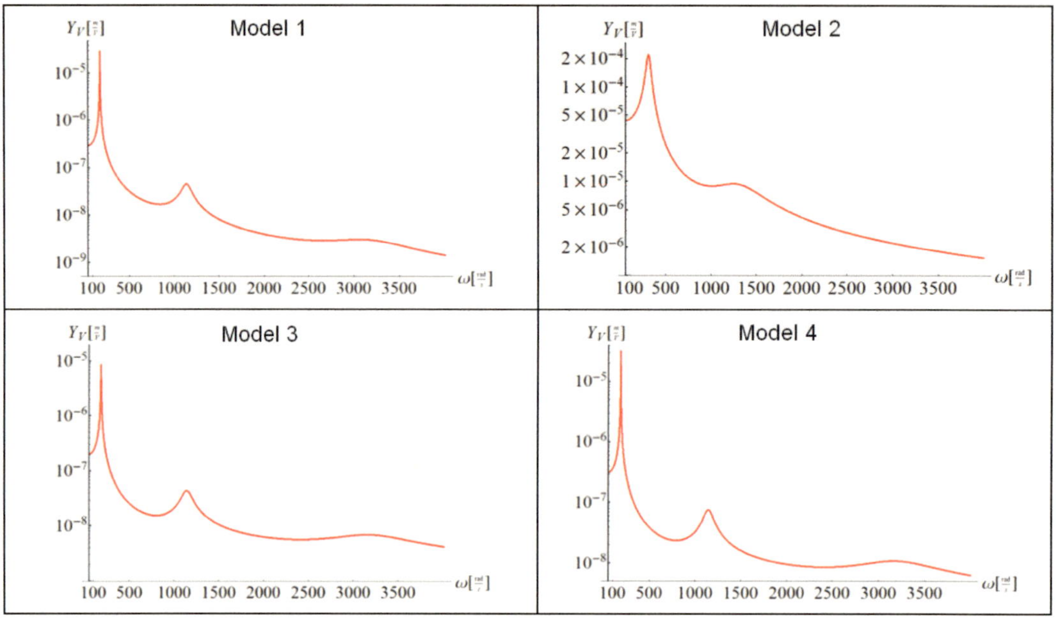

Fig. 14. Absolute value of the dynamic characteristic of mechatronic system with piezoelectric actuator, for the first three natural frequencies (a half logarithmic scale)

Dimensionless frequencyies of external force or electric voltage were introduced by dividing their values by the value of the first natural frequency of the mechanical subsystem. Dimensionless values of analyzed parameters were obtained by dividing them by their initial values. Obtained results for selected parameters are presented in Fig. 15 and Fig. 16.

Influence of the other parameters on characteristics of considered systems were analyzed in other publications (Buchacz & Płaczek, 2009b, 2010a).

7. Conclusions and selection of an optimal mathematical model

Realized studies have shown that the corrected approximate Galerkin method can be used to analyze mechatronic systems with piezoelectric transducers. Verification of the approximate method proved that obtained results can be treated as very precise. Precision of the mathematical model of considered system has no big influence on the final results. There are no significant differences between the values of natural vibration frequencies of considered systems and course of dynamic characteristics, except the second model. In case of the mathematical model with the assumption of pure shear of the glue layer a very significant shift of natural frequencies values and increase of piezoelectric damper or actuator efficiency were observed. These discrepancies are the results of the assumed

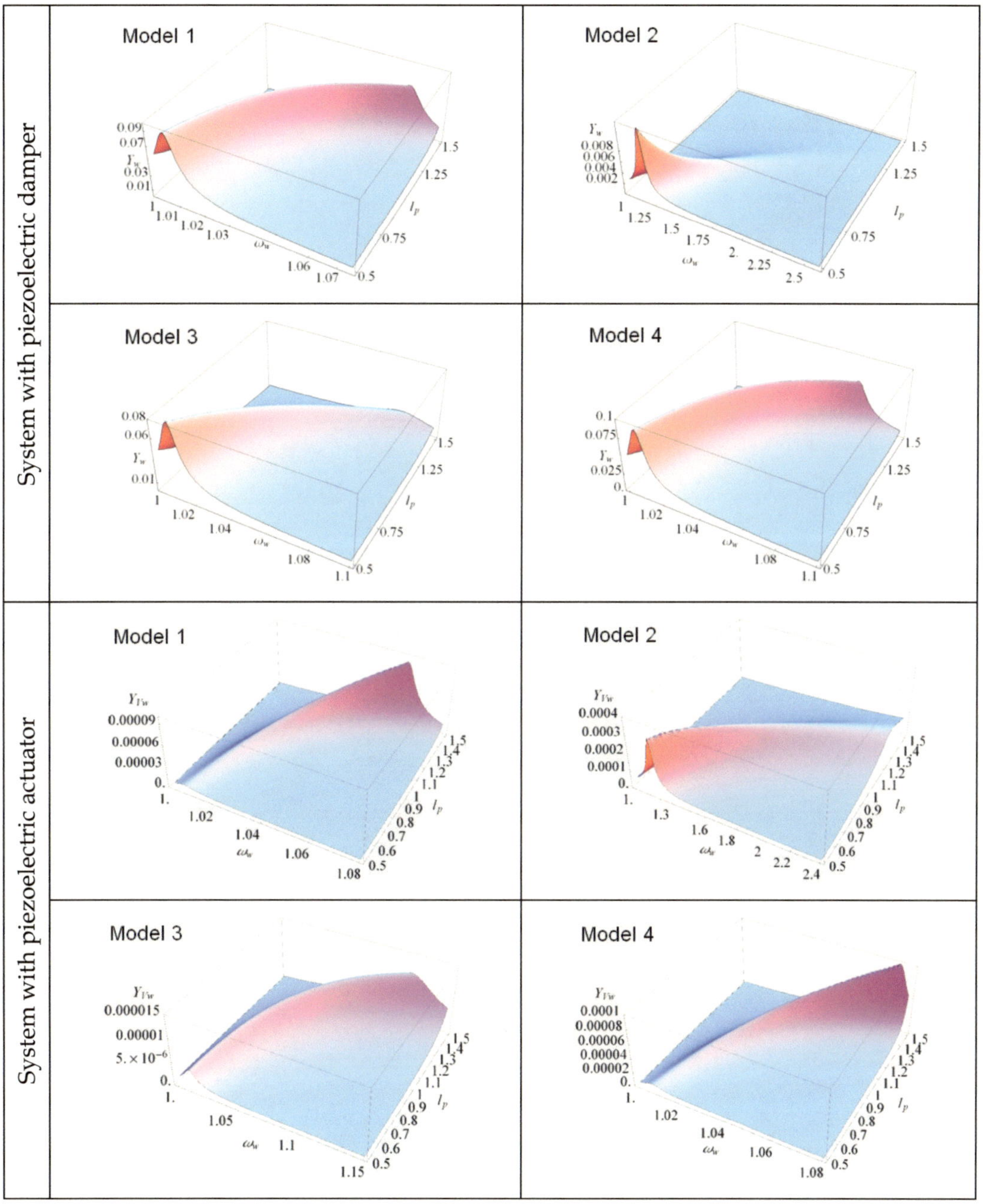

Fig. 15. Influence of length of the piezoelectric transducer on the absolute value of the dimensionless dynamic characteristics

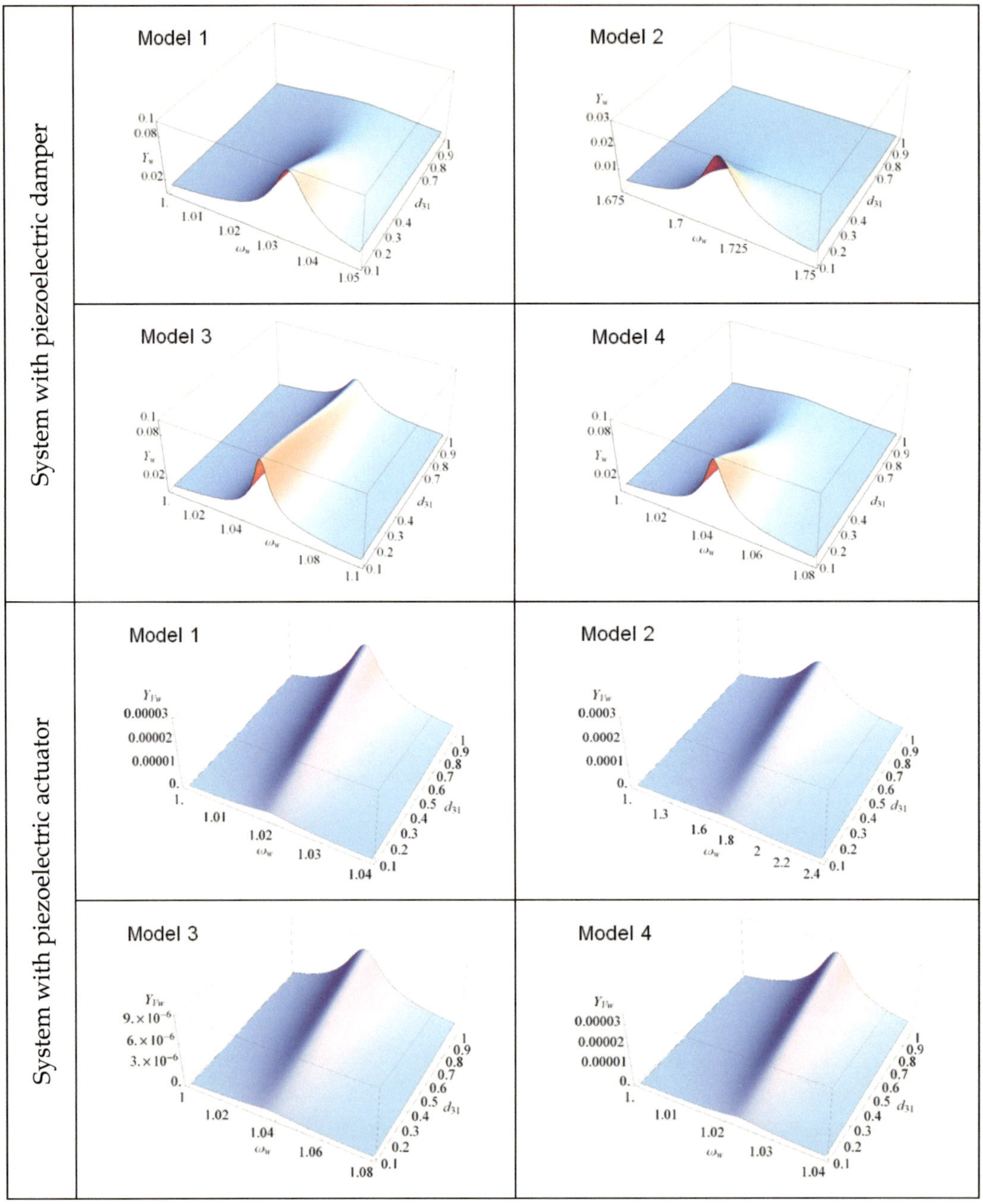

Fig. 16. Influence of piezoelectric constant of the piezoelectric transducer on the absolute value of the dimensionless dynamic characteristics

simplifications of the real strain of the transducer and resulting generated shear stress in the glue layer. There was also an assumption about pure shear of the glue layer, while, in the real system, this layer is under the influence of forces that cause the eccentric tension of it.

The simplest is the mathematical model with the assumption about perfectly bonded piezoelectric transducer. But taking this assumption it is impossible to define influence of the glue layer on the dynamic characteristic of the system. Using this model it is not possible to meet requirements undertaken in this work. To take into account properties of the glue layer and its real loads to which it is subjected, mathematical models, where an eccentric tension of glue layer was considered, were developed. Interactions between elements of the system were being taken into consideration and real strain of the transducer was determined. The third mathematical model is much more complex then the last one, while obtained results are very similar. It is therefore concluded that the optima, in terms of assumed criteria, is the last mathematical model where a bending moment generated by the transducer and eccentric tension of a glue layer between the piezoelectric transducer and surface of the beam were taken into account. Using this model it is possible to analyze influence of all components of the system, including glue layer between the beam and transducer, while it is quite simple at the same time.

8. References

Behrens S., Fleming A. J., & Moheimani S. O. R. (2003). A broadband controller for shunt piezoelectric damping of structural vibration. *Smart Materials and Structures*, Vol. 12, (2003), pp. 18-28

Buchacz A., & Płaczek M. (2009a). Damping of Mechanical Vibrations Using Piezoelements, Including Influence of Connection Layer's Properties on the Dynamic Characteristic. *Solid State Phenomena*, Vols. 147-149, (2009), pp. 869-875, ISSN: 1012-0394

Buchacz A., & Płaczek M. (2009b). The discrete-continuous model of the one-dimension vibrating mechatronic system. *PAMM - Proc. Appl. Math. Mech.*, Vol. 9, No. 1, (2009), pp. 395-396, ISSN: 1617-7061

Buchacz A., & Płaczek M. (2009c). The vibrating mechatronic system modeled as the combined beam, *Proceedings of International Scientific and Technical Conference Reliability and Durability of Mechanic and Biomechanical Systems and Elements of their Constructions*, Sevastopol, September 2009

Buchacz A., & Płaczek M. (2010a). Selection of Parameters of External Electric Circuit for Control of Dynamic Flexibility of a Mechatronic System. *Solid State Phenomena*, Vol. 164, (2010), pp. 323-326, ISSN: 1012-0394

Buchacz A., & Płaczek M. (2010b). Development of Mathematical Model of a Mechatronic System. *Solid State Phenomena*, Vol. 164, (2010), pp. 319-322, ISSN: 1012-0394

Buchacz A., & Płaczek M. (2010c). The analysis of vibrating systems based on the exact end approximate method. *International Journal of Modern Manufacturing Technologies*, Vol. II, No. 1, (2010), pp. 19-24, ISSN: 2067-3604

Buchacz A., & Płaczek M. (2010d). The exact and approximate method in mechanical system's analysis, *PAMM - Proc. Appl. Math. Mech.*, Vol. 10, No. 1, (2010), pp. 379-380, ISSN: 1617-7061

Buchacz A., & Płaczek M. (2011). Characteristic of the mechatronic system with piezoelectric actuator modeled as the combined beam, *Proceedings of The 15th International Conference Modern Technologies, Quality and Innovation ModTech 2011*, ISSN: 2069-6736, Vadul lui Voda, Chisinau, Republic of Moldova, May 2011

Fein O. M. (2008). A model for piezo-resistive damping of two-dimensional structures. *Journal of Sound and Vibration*, Vol. 310, No. 4-5, (2008), pp. 865-880, ISSN: 0022-460X

Fleming A. J., Behrens S., & Reza Moheimani S. O. (2002). Optimization and Implementation of Multimode Piezoelectric Shunt Damping Systems. *IEEE/ASME Transactions on Mechatronics*, Vol. 7, No. 1, (March 2002), pp. 87-94.

Gao J. X., & Liao W. H. (2005). Vibration analysis of simply supported beams with enhanced self-sensing active constrained layer damping treatments. *Journal of Sound and Vibration*, Vol. 280, (2005), pp. 329-357

Hagood N. W., & von Flotow A. (1991). Damping of structural vibrations with piezoelectric materials and passive electric networks. *Journal of Sound and Vibration*, Vol. 146, No. 2, (1991), pp. 243-268

Kurnik W., Przybyłowicz P.M., & Tylikowski A. (1995). Torsional Vibrations Actively Attenuated by Piezoelectric System, *Proceedings of the 4th German-Polish Workshop on Dynamical Problems in Mechanical Systems*, Berlin, July 1995

Kurnik W. (2004). Damping of Mechanical Vibrations Utilising shunted Piezoelements. *Machine Dynamics Problems*, Vol. 28, No. 4, (2004), pp. 15-26

Moheimani S.O.R., & Fleming A.J. (2006). *Piezoelectric Transducers for Vibration control and Damping*, Springer, ISBN: 1-84628-331-0, London

Pietrzakowski M. (2001). Active damping of beams by piezoelectric system: effects of bonding layer properties. *International Journal of Solids and Structures*, Vol. 38, (2001), pp. 7885-7897, ISSN: 0020-7683

Preumont A. (2006). *Mechatronics: Dynamics of Electromechanical and Piezoelectric Systems*, Springer, ISBN: 1402046952, Dordrecht, The Netherlands

Yoshikawa S., Bogue A., & Degon B. (1998). Commercial Application of Passive and Active Piezoelectric Vibration Control, *Proceedings of the Eleventh IEEE International Symposium on Applications of Ferroelectrics*, ISBN: 0-7803-4959-8, Montreux, Switzerland, August 1998

3

Distributed-Parameter Modeling of Energy Harvesting Structures with Discontinuities

Adam Wickenheiser
George Washington University,
United States

1. Introduction

Energy harvesting - the ability to gather energy from the local environment to power wireless devices - has seen significant development over the past decade as the demand for portable electronics increases. Although on-board batteries provide a simple means of providing energy for these devices, their energy density can be insufficient for miniature devices or long-term deployment (Anton & Sodano, 2007). A means of replenishing on-board energy storage has the potential to reduce the frequency of battery replacement or eliminate the need altogether. Vibration-based energy harvesting in particular has garnered much attention due to the ubiquity of vibrational energy in the environment (Roundy et al., 2003). Several methods of electromechanical transduction from vibrations have been investigated, including electromagnetic induction, electrostatic varactance, and the piezoelectric effect, the latter being the province of this chapter.

Mechanical energy is transformed into electricity by straining piezoelectric material mounted on a structure that is subjected to ambient vibrations. If a natural frequency of the structure is matched to the predominant excitation frequency, resonance occurs, where large strains are induced by relatively small excitations. A major problem with resonant vibration-based energy harvesters is that their peak strain (and hence, power) only occurs near the natural frequencies of the transducer. For many potential applications, ambient vibrations are low frequency, requiring large or heavy structures for resonance (Roundy et al., 2003; Wickenheiser & Garcia, 2010a). In order to shrink the size and mass of these devices while reducing their natural frequencies, a variety of techniques have been employed. For example, changing the standard cantilevered beam geometry and manipulating the mass distribution along the beam have been investigated. Varying the cross sections along the beam length (Dietl & Garcia, 2010; Reissman et al., 2007; Roundy et al., 2005) and the ratio of tip mass to beam mass (Dietl & Garcia, 2010; Wickenheiser, 2011) have been shown to improve the electromechanical coupling (a factor in the energy conversion rate) over a uniform cantilever beam design. Changing the number and location of piezoelectric patches or layers along the beam can improve coupling and shift the natural frequency of the device (Guyomar et al., 2005; Wu et al., 2009). Multi-beam structures can compact the design by folding it in on itself while retaining a similar natural frequency to the original, straight configuration (Karami & Inman, 2011; Erturk et al., 2009). A nonlinear technique called "frequency up-conversion" also shows promise to boost power at frequencies more than an order of magnitude below resonance (Murray & Rastegar, 2009; Tieck et al., 2006;

Wickenheiser & Garcia, 2010b). Despite the prevalence of widely varying designs, no single analytic method exists for predicting the electromechanical behavior of these systems.

In the energy harvesting literature, the piezoelectric transducer is commonly modeled as a lumped, single-degree-of-freedom (DOF) system, typically a current source in parallel with an intrinsic capacitance. To more accurately predict the dynamics of energy harvesters, mechanical models have been developed based on their geometry and material properties. Two common approaches to modeling and simulating these devices are lumped parameter (typically single DOF) (duToit et al., 2005; Roundy & Wright, 2004) and distributed parameter (multi-DOF) (duToit et al., Erturk & Inman, 2008; Sodano et al., 2004; Wickenheiser & Garcia, 2010c) models. Lumped parameter models are simple and accurate when vibrating near a resonant frequency and experimental data are available to estimate the model parameters. Distributed parameter models are more accurate when multiple modes of vibration are expressed, can predict geometric effects such as charge cancellation, and can be easily extended to include arbitrary DOFs. However, these models are much more complex, are designed for a specific geometry, and require experimental determination of some of their parameters.

In this chapter, a straightforward analytic approach is taken for modeling beams of varying cross-sectional geometry and multiple discontinuities, including lumped masses and bends. This technique also correctly accounts for the changes in the mechanical response from adding piezoelectric layers with partial coverage to the structure. This method is derived from the classical transfer matrix method (TMM) for multi-component structures and trusses (Pestel & Leckie, 1963) combined with an existing model for constant cross section, Euler-Bernoulli beam energy harvesters (Wickenheiser & Garcia, 2010c). A variation of this technique is employed by (Karami & Inman, 2011) to find the natural frequencies and mode shapes of a zigzag structure; however, their formulation is specific to 180° bends between segments. The TMM has been shown to reduce to the classical solutions (e.g. cantilevered beam with or without tip mass) for structures consisting of a single segment (Reissman et al., 2011). An advantage of this method is that increasing the complexity of the structure does not increase the size of the eigenvalue problem required to find the natural frequencies and mode shapes. Furthermore, the same formulation can be used for an arbitrary distribution of lumped masses, bends between members, and varying geometry beam segments.

In the following sections, the equations of motion (EOMs) are derived for uniform beam segments and for lumped masses. Subsequently, it is shown how these subsystems can be combined to form an arbitrarily complex structure. The eigenvalue problem for this class of design is then solved for the natural frequencies and mode shapes. These solutions are incorporated into a partial differential equation (PDE) model that includes the linearized piezoelectric constitutive equations, enabling the solution of the coupled electromechanical dynamics. Finally, a few simple case studies are presented to highlight the usefulness of this technique.

2. Derivation of TMM for Euler-Bernoulli beam structures

2.1 Overview of methodology

The transfer matrix method used in this study is derived from the methodology described in (Pestel & Leckie, 1963). This method is used to calculate the natural frequencies and mode shapes (i.e. the eigensolution) for piecewise continuous structures, such as the one shown in

Fig. 1. This figure shows a 3-segment beam with lumped masses connected to the tip of each segment. (In this discussion, the "base" of each segment is the end closest to the host structure, whereas the "tip" is the end furthest.) Each segment is assumed to have constant geometric and material properties; however, different segments may have different properties. The lumped masses and bend angles can vary for each segment, including the case of no lumped mass between two segments. Furthermore, each segment may have a different number and arrangement of piezoelectric and substructure layering - e.g. combining bimorph, unimorph, and bare substructure.

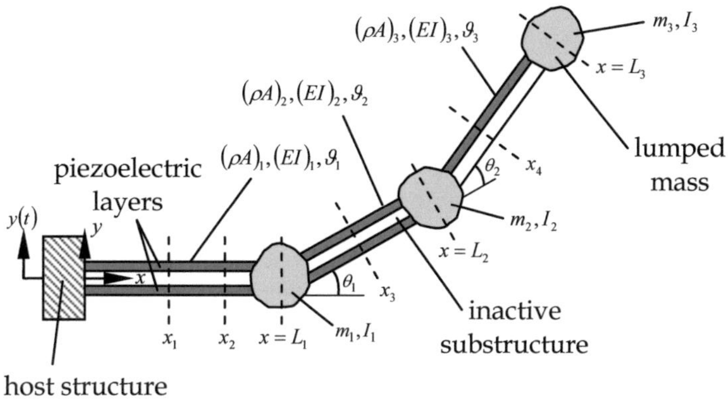

Fig. 1. Layout and geometric parameters of an example piecewise continuous structure.

Let $w(x,t)$ be the deflection of the beam in the transverse direction and $u(x,t)$ be the deflection in the axial direction, each measured relative to the equilibrium position of the structure. Separation of variables is adopted to decompose these deflections into spatial and temporal components:

$$w(x,t) = \sum_{r=1}^{\infty} \eta_r(t)\phi_r(x) \text{ and } u(x,t) = \sum_{r=1}^{\infty} \eta_r(t)\psi_r(x) \tag{1}$$

where $\eta_r(t)$ is the rth modal displacement, $\phi_r(x)$ is the rth transverse mode shape function, and $\psi_r(x)$ is the rth axial mode shape function. The subscript r is henceforth dropped for clarity, since the following discussion applies to any mode.

As will be discussed in the following sections, Euler-Bernoulli beam theory requires 4 states to describe the variation of ϕ with respect to x, namely the mode shape itself ϕ, its slope $d\phi/dx$, the internal bending moment M, and the internal shear force V. The state equation for the variation of ψ with respect to x includes the mode shape itself ψ and the normal (i.e. axial force) N. Assembling these variables into a state vector

$$\mathbf{z} = \begin{bmatrix} \psi & N & \phi & \frac{d\phi}{dx} & M & V \end{bmatrix}^T \tag{2}$$

a 6x6 linear system of the form

$$\frac{d\mathbf{z}}{dx} = \mathbf{A}(x)\mathbf{z}(x) \tag{3}$$

will be derived subsequently. Using the state transition matrix $\mathbf{\Phi}$ of Eq. (3), the state vectors at any two points along the structure can be related using

$$\mathbf{z}(x_2) = \mathbf{\Phi}(x_2, x_1)\mathbf{z}(x_1) \tag{4}$$

At this stage, the power of the transfer matrix method becomes apparent. Consider the problem of relating the states (components of $\mathbf{z}$) between points x_1 and x_2 and between points x_3 and x_4 shown in Fig. 1. In the next sections, state transition matrices will be derived for each beam segment, called *field transfer matrices*, and each lumped mass, called *point transfer matrices*. Denoting the field transfer matrix for the jth segment $\mathbf{F_j}$ and the point transfer matrix for the jth lumped mass $\mathbf{P_j}$, it will be shown that Eq. (4) can be written as

$$\mathbf{z}(x_2) = \mathbf{F_1}(x_2 - x_1)\mathbf{z}(x_1) \tag{5a}$$

between points x_1 and x_2 and

$$\mathbf{z}(x_4) = \mathbf{F_3}(x_4 - L_2)\mathbf{P_2}\mathbf{F_2}(L_2 - x_3)\mathbf{z}(x_3) \tag{5b}$$

between points x_3 and x_4, using the semigroup property of state transition matrices. Eq. (5b) also displays another feature of the transfer matrix method: no matter how many beam segments and lumped masses there are in the structure, the problem never grows beyond a 6x6 linear system.

2.2 Derivation of EOMs for an Euler-Bernoulli beam segment

In this section, the EOMs for the states across a uniform beam segment are derived using Euler-Bernoulli beam assumptions and linearized material constitutive equations. The approach taken herein is based on force and moment balances and is a generalization of the treatments by (Erturk & Inman, 2008; Söderkvist, 1990; Wickenheiser & Garcia, 2010c). It is assumed that each beam segment is uniform in cross section and material properties. Furthermore, the standard Euler-Bernoulli beam assumptions are adopted, including negligible rotary inertia and shear deformation (Inman, 2007).

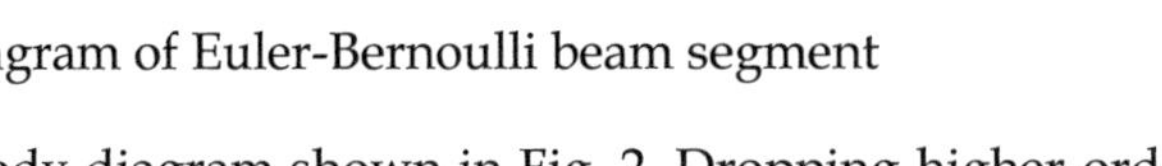

Fig. 2. Free-body diagram of Euler-Bernoulli beam segment

Consider the free-body diagram shown in Fig. 2. Dropping higher order terms, balances of forces in the y-direction and moments yield

$$\frac{\partial V(x,t)}{\partial x} + f(x,t) = (\rho A)\frac{\partial^2 w(x,t)}{\partial t^2} \tag{6a}$$

$$\frac{\partial M(x,t)}{\partial x} = -V(x,t) \tag{6b}$$

where $V(x,t)$ is the shear force, $M(x,t)$ is the internal moment generated by mechanical and electrical strain, $f(x,t)$ is the externally applied force per unit length (it will be shown later that this is the inertial force induced by the base excitation), and (ρA) is the mass per unit length (Inman, 2007). Note that if the segment is monolithic, (ρA) is simply the product of the density of the material and the cross-sectional area. For the case of a bimorph beam segment, this term is given by

$$(\rho A) = \frac{m}{l} = \frac{\rho_s t_s bl + 2\rho_p t_p bl}{l} = b\left(\rho_s t_s + 2\rho_p t_p\right) \tag{7}$$

The internal bending moment is the net contribution of the stresses in the axial direction in the beam. The stress within the piezoelectric layers is found from the linearized constitutive equations

$$\begin{aligned} T_1 &= c_{11}^E S_1 - e_{31} E_3 \\ D_3 &= e_{31} S_1 + \varepsilon_{33}^S E_3 \end{aligned} \tag{8}$$

where T is stress, S is strain, E is electric field, D is electric displacement, c is Young's Modulus, e is piezoelectric constant, and ε is dielectric constant. The subscripts indicate the direction of perturbation; in the cantilever configuration shown in Fig. 1, 1 corresponds to axial and 3 corresponds to transverse. The superscript $(\cdot)^E$ indicates a linearization at constant electric field, and the superscript $(\cdot)^S$ indicates a linearization at constant strain (IEEE, 1987). The stress within the substrate layer(s) is given simply by the linear stress-strain relationship $T_1 = c_{11,s} S_1$, where $c_{11,s}$ is Young's Modulus of the substrate material in the axial direction. Since deformations are assumed small, the axial strain is the same as the case of pure bending, which is given by $S_1 = -y\,\partial^2 w(x,t)/\partial x^2$ (Beer & Johnson, 1992), and the transverse electric field is assumed constant and equal to $E_3 = \pm v(t)/t_p$, where $v(t)$ is the voltage across the electrodes, and the top and bottom layer have opposite signs due to the parallel configuration wiring. (This approximation is reasonable given the thinness of the layers.) Consider the case of a bimorph beam segment of width b, substrate layer thickness t_s, and piezoelectric layer thickness t_p. Then the bending moment is

$$\begin{aligned} M(x,t) = & \int_{-t_s/2-t_p}^{-t_s/2} T_1 by\,dy + \int_{-t_s/2}^{t_s/2} T_1 by\,dy + \int_{t_s/2}^{t_s/2+t_p} T_1 by\,dy \\ = & -\left[\int_{-t_s/2-t_p}^{-t_s/2} c_{11}^E by^2 dy + \int_{-t_s/2}^{t_s/2} c_{11,s} by^2 dy + \int_{t_s/2}^{t_s/2+t_p} c_{11}^E by^2 dy\right]\frac{\partial^2 w(x,t)}{\partial x^2} \\ & -\left[\int_{-t_s/2-t_p}^{-t_s/2} \frac{e_{31}}{t_p} by\,dy - \int_{t_s/2}^{t_s/2+t_p} \frac{e_{31}}{t_p} by\,dy\right] v(t)\left[H(x-L_L) - H(x-L_R)\right] \\ = & \underbrace{\left\{c_{11,s} b\frac{t_s^3}{12} + 2c_{11}^E b\left[\frac{t_p^3}{12} + t_p\left(\frac{t_p + t_s}{2}\right)^2\right]\right\}}_{(EI)}\frac{\partial^2 w(x,t)}{\partial x^2} + \underbrace{-e_{31} b\left(t_s + t_p\right)}_{\vartheta} v(t)\left[H(x-L_L) - H(x-L_R)\right] \end{aligned} \tag{9}$$

where $H(\cdot)$ is the Heaviside step function, and L_L, L_R are the left and right ends of the segment, respectively. In Eq. (9), the constant multiplying the $\partial^2 w(x,t)/\partial x^2$ term is defined as (EI), the effective bending stiffness. (Note that if the beam segment is monolithic, this constant is simply the product of the Young's Modulus and the moment of inertia.) The constant multiplying the $v(t)$ term is defined as ϑ, the electromechanical coupling coefficient. Substituting Eq. (9) into Eq. (6) yields

$$(\rho A)\frac{\partial^2 w(x,t)}{\partial t^2} + (EI)\frac{\partial^4 w(x,t)}{\partial x^4} + \vartheta\left[\frac{d\delta(x-L_L)}{dx} - \frac{d\delta(x-L_R)}{dx}\right]v(t) = f(x,t) \tag{10}$$

which is the transverse mechanical EOM for a beam segment.

The electrical EOM can be found by integrating the electric displacement over the surface of the electrodes, yielding the net charge $q(t)$ (IEEE, 1987):

$$\begin{aligned} q(t) &= \iint_{\substack{\text{upper}\\ \text{layer}}} D_3 dA - \iint_{\substack{\text{lower}\\ \text{layer}}} D_3 dA \\ &= b\int_{L_L}^{L_R}\left[\frac{1}{t_p}\int_{t_s/2}^{t_s/2+t_p} -e_{31} y \frac{\partial^2 w(x,t)}{\partial x^2} dy - \frac{\varepsilon_{33}^S}{t_p} v(t)\right] dx \\ &\quad - b\int_{L_L}^{L_R}\left[\frac{1}{t_p}\int_{-t_s/2-t_p}^{-t_s/2} -e_{31} y \frac{\partial^2 w(x,t)}{\partial x^2} dy + \frac{\varepsilon_{33}^S}{t_p} v(t)\right] dx \\ &= \underbrace{-e_{31} b\left(t_s + t_p\right)}_{\vartheta}\left[\left.\frac{\partial w(x,t)}{\partial x}\right|_{x=L_R} - \left.\frac{\partial w(x,t)}{\partial x}\right|_{x=L_L}\right] - \underbrace{\frac{2\varepsilon_{33}^S bL}{t_p}}_{C} v(t) \end{aligned} \tag{11}$$

where the constant multiplying the $v(t)$ term is defined as C, the net clamped capacitance of the segment. Eqs. (10–11) provide a coupled system of equations; these can be solved by relating the voltage $v(t)$ to the charge $q(t)$ through the external electronic interface.

To derive the EOMs for the axial motion of each segment, it is assumed that the deformations in this direction are negligible compared to the transverse deformations. This assumption is reasonable if the cross sections are very thin in the transverse direction, in which case $A >> I$. Thus, if the beam is assumed rigid in the *x*-direction, a balance of forces gives

$$(\rho A)\frac{\partial^2 u(x,t)}{\partial t^2} + \frac{\partial N(x,t)}{\partial x} = 0 \tag{12}$$

which constitutes the EOM for the axial direction for each beam segment. It should be noted that in Eqs. (10–12), the constants in the equations have been derived for bimorph segments; constants for other configurations can be found in (Wickenheiser & Garcia, 2010c). These three equations are the EOMs for this structure, which are solved in Section 4.

2.3 Field transfer matrix derivation

To derive the state transition matrix between two points along a uniform beam segment, the Euler-Bernoulli EOMs derived in the previous section are employed, dropping the electromechanical coupling effects and the inertial forces due to base excitation, i.e. setting $v(t)\equiv 0$ and $f(x,t)\equiv 0$. This is equivalent to the assumption of Euler-Bernoulli mode shapes when modeling piezoelectric benders, a prevalent simplification appearing in the literature (duToit et al., 2005; Erturk & Inman, 2008; Wickenheiser & Garcia, 2010c). Under these assumptions, Eqs. (6,9,12) become

$$\frac{\partial V(x,t)}{\partial x}=(\rho A)_j\frac{\partial^2 w(x,t)}{\partial t^2},\quad \frac{\partial M(x,t)}{\partial x}=-V(x,t),$$

$$(\rho A)_j\frac{\partial^2 u(x,t)}{\partial t^2}+\frac{\partial N(x,t)}{\partial x}=0 \text{ and } M(x,t)=(EI)_j\frac{\partial^2 w(x,t)}{\partial x^2} \tag{13}$$

for beam segment j.

At this point, Eq. (1) is applied. Each mode shape has a natural frequency ω associated with it (dropping the r subscript). With this substitution, the first and third of the previous equations can be rewritten as

$$\frac{dV(x)}{dx}=-(\rho A)_j\omega^2\phi(x) \text{ and } \frac{dN(x)}{dx}=(\rho A)_j\omega^2\psi(x) \tag{14}$$

Collecting Eqs. (13–14) and writing them in terms of the mode shapes yields the linear system

$$\frac{d}{dx}\underbrace{\begin{bmatrix}\psi\\ N\\ \phi\\ d\phi/dx\\ M\\ V\end{bmatrix}}_{\mathbf{z}}=\underbrace{\left[\begin{array}{cc|cccc}0&0&0&0&0&0\\ (\rho A)_j\omega^2&0&0&0&0&0\\ \hline 0&0&0&1&0&0\\ 0&0&0&0&\frac{1}{(EI)_j}&0\\ 0&0&0&0&0&-1\\ 0&0&-(\rho A)_j\omega^2&0&0&0\end{array}\right]}_{\mathbf{A}_\mathrm{j}}\underbrace{\begin{bmatrix}\psi\\ N\\ \phi\\ d\phi/dx\\ M\\ V\end{bmatrix}}_{\mathbf{z}} \tag{15}$$

which is the form sought in Eq. (3). Note that the transverse and axial dynamics are decoupled.

Within a beam segment, the cross sections are assumed constant along the length, which has resulted in a constant state matrix $\mathbf{A}_\mathrm{j}$ in Eq. (15). Hence, from linear systems theory, the state transition matrix is simply a function of the *difference* in the positions along the beam, i.e. $\mathbf{\Phi}(x_2,x_1)=\mathbf{\Phi}(x_2-x_1)\equiv\mathbf{\Phi}(\Delta x)$. Thus, the field transfer matrix for beam segment j can be written as $\mathbf{F}_\mathrm{j}(\Delta x)=e^{\mathbf{A}_\mathrm{j}\Delta x}$.

Since $\mathbf{A}_\mathrm{j}$ is block diagonal, the matrix exponential can be computed for each block separately. The upper left block can be integrated explicitly. An analytical formula for the matrix exponential of the lower-right block, labeled $\mathbf{B}_\mathrm{j}$, can be found using the Cayley-Hamilton theorem, which states

$$e^{\mathbf{B}_j\Delta x} = c_0 I + c_1\left(\mathbf{B}_j\Delta x\right) + c_2\left(\mathbf{B}_j\Delta x\right)^2 + c_3\left(\mathbf{B}_j\Delta x\right)^3 \tag{16}$$

This equation must hold when $\mathbf{B}_j$ is replaced by any of its eigenvalues, which are given by $\lambda = \pm\beta$ and $\lambda = \pm i\beta$, where

$$\beta^4 = \frac{(\rho A)_j \omega^2}{(EI)_j} \tag{17}$$

Substituting these eigenvalues into Eq. (16) yields a system of 4 equations for the unknowns $c_0, \ldots, c_3$. The solution of these equations is

$$\begin{aligned} c_0 &= \frac{1}{2}\left[\cosh(\beta\Delta x) + \cos(\beta\Delta x)\right] \\ c_1 &= \frac{1}{2(\beta\Delta x)}\left[\sinh(\beta\Delta x) + \sin(\beta\Delta x)\right] \\ c_2 &= \frac{1}{2(\beta\Delta x)^2}\left[\cosh(\beta\Delta x) - \cos(\beta\Delta x)\right] \\ c_3 &= \frac{1}{2(\beta\Delta x)^3}\left[\sinh(\beta\Delta x) - \sin(\beta\Delta x)\right] \end{aligned} \tag{18}$$

Substituting these formulas back into Eq. (16) and concatenating with the upper-left block yields

$$\mathbf{F}_j(\Delta x) = \left[\begin{array}{cc|cccc} 1 & 0 & 0 & 0 & 0 & 0 \\ \Delta x(\rho A)_j \omega^2 & 1 & 0 & 0 & 0 & 0 \\ \hline 0 & 0 & c_0 & \Delta x c_1 & \frac{(\Delta x)^2}{(EI)_j} c_2 & -\frac{(\Delta x)^3}{(EI)_j} c_3 \\ 0 & 0 & \frac{(\Delta x)^3 (\rho A)_j \omega^2}{(EI)_j} c_3 & c_0 & \frac{\Delta x}{(EI)_j} c_1 & -\frac{(\Delta x)^2}{(EI)_j} c_2 \\ 0 & 0 & (\Delta x)^2 (\rho A)_j \omega^2 c_2 & (\Delta x)^3 (\rho A)_j \omega^2 c_3 & c_0 & -\Delta x c_1 \\ 0 & 0 & -\Delta x (\rho A)_j \omega^2 c_1 & -(\Delta x)^2 (\rho A)_j \omega^2 c_2 & -\frac{(\Delta x)^3 (\rho A)_j \omega^2}{(EI)_j} c_3 & c_0 \end{array}\right] \tag{19}$$

Eq. (19) is the field transfer matrix of a beam section for relating the state vectors $\mathbf{z}$ at different positions within a single beam segment. A use of this matrix for that purpose is seen in Eq. (5a).

2.4 Point transfer matrix derivation

The point transition matrix $\mathbf{P}$ is now derived, which accounts for discontinuities between the uniform beam segments. Consider the free-body diagram of the lumped mass shown in Fig. 2. This mass is considered a point mass with mass m_j and rotary inertia I_j, located at $x = L_j$. Since the mass is assumed to be infinitesimal in size, the forces and moments are evaluated at $x = L_j -$ and $x = L_j +$, meaning approaching $x = L_j$ from the left and the right, respectively.

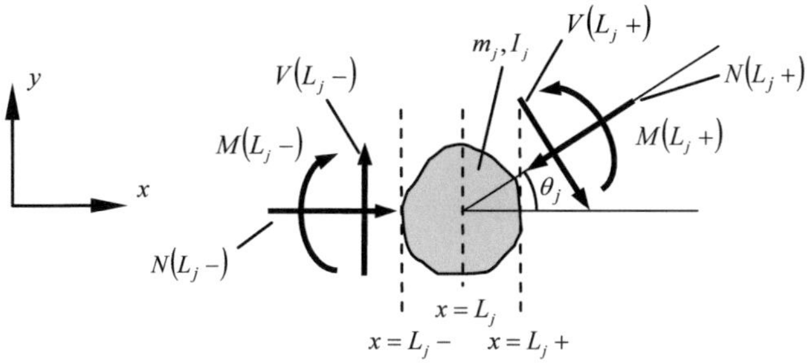

Fig. 3. Forces and moments on a lumped mass located at $x = L_j$.

The slope of the beam is continuous across the lumped mass, hence $d\phi(L_j -)/dx = d\phi(L_j +)/dx$. However, due to the rotation of the local beam coordinate system from one side of the lumped mass to the other, the mode shapes are not continuous, i.e.

$$\begin{bmatrix} \psi(L_j +) \\ \phi(L_j +) \end{bmatrix} = \begin{bmatrix} \cos\theta_j & \sin\theta_j \\ -\sin\theta_j & \cos\theta_j \end{bmatrix} \begin{bmatrix} \psi(L_j -) \\ \phi(L_j -) \end{bmatrix} \tag{20}$$

Furthermore, due to the lumped inertia, the shear force, normal force, and bending moment are not continuous. A balance of forces and moments on the lumped mass, referring to Fig. 3, gives

$$N(L_j +) = (\cos\theta_j) N(L_j -) + (\sin\theta_j) V(L_j -) + (m_j\omega^2 \cos\theta_j)\psi(L_j -) + (m_j\omega^2 \sin\theta_j)\phi(L_j -) \tag{21}$$

$$V(L_j +) = (-\sin\theta_j) N(L_j -) + (\cos\theta_j) V(L_j -) + (-m_j\omega^2 \sin\theta_j)\psi(L_j -) + (m_j\omega^2 \cos\theta_j)\phi(L_j -) \tag{22}$$

$$M(L_j +) = -I_j\omega^2 \frac{d\phi(L_j -)}{dx} + M(L_j -) \tag{23}$$

Assembling these equations together yields

$$\underbrace{\begin{bmatrix} \psi(L_j +) \\ N(L_j +) \\ \phi(L_j +) \\ d\phi(L_j +)/dx \\ M(L_j +) \\ V(L_j +) \end{bmatrix}}_{\mathbf{z}(L_j +)} = \underbrace{\begin{bmatrix} \cos\theta_j & 0 & \sin\theta_j & 0 & 0 & 0 \\ m_j\omega^2\cos\theta_j & \cos\theta_j & m_j\omega^2\sin\theta_j & 0 & 0 & \sin\theta_j \\ -\sin\theta_j & 0 & \cos\theta_j & 0 & 0 & 0 \\ 0 & 0 & 0 & 1 & 0 & 0 \\ 0 & 0 & 0 & -I_j\omega^2 & 1 & 0 \\ -m_j\omega^2\sin\theta_j & -\sin\theta_j & m_j\omega^2\cos\theta_j & 0 & 0 & \cos\theta_j \end{bmatrix}}_{\mathbf{P_j}} \underbrace{\begin{bmatrix} \psi(L_j -) \\ N(L_j -) \\ \phi(L_j -) \\ d\phi(L_j -)/dx \\ M(L_j -) \\ V(L_j -) \end{bmatrix}}_{\mathbf{z}(L_j -)} \tag{24}$$

which provides a formula for the point transition matrix $\mathbf{P_j}$ of the jth lumped mass. This formula is valid when the lumped mass is at the tip of the structure, in which case $M(L_j+)=V(L_j+)=N(L_j+)=0$ in Eq. (24) (i.e. the free end condition), or if there is no lumped mass between two beam segments, a situation given as a case study below. In this latter case, $m_j=I_j=0$ in Eq. (16). If, furthermore, there is no angle between beam segments, i.e. $\theta_j=0$, then $\mathbf{P_j}$ reduces to the identity matrix, indicating that all of the states are continuous through the junction.

3. Eigensolution using the system transfer matrix

3.1 Natural frequencies

As discussed in section 2.1, the state transition matrix $\mathbf{\Phi}(x_2,x_1)$ relates the states of the system between any points along the beam through Eq. (2). Depending on the locations of x_1 and x_2, the transition matrix is, in general, expressible as a product of field and point transfer matrices, as illustrated by Eqs. (5a–b). The number of matrices in this product is equal to the number of beam segments and junctions between the two points.

It should be noted, though, that at this point the natural frequency ω is still unknown; thus, $\mathbf{\Phi}(x_2,x_1)$ cannot be evaluated between any two points in general. However, the boundary conditions at the ends of the structure provide locations where some of the states are known. In the presently studied cantilever (or "fixed-free") configuration, the following states are known:

$$\psi(0)=\phi(0)=\frac{d\phi}{dx}(0)=0 \text{ and } N(L_n)=M(L_n)=V(L_n)=0 \tag{25}$$

where n is the total number of beam segments. These boundary conditions signify a fixed condition at $x=0$ and a free condition at $x=L_n$. To relate the fixed and free ends, Eq. (4) is employed:

$$\begin{bmatrix} \psi(L_n) \\ N(L_n) \\ \phi(L_n) \\ d\phi(L_n)/dx \\ M(L_n) \\ V(L_n) \end{bmatrix} = \underbrace{\left(\prod_{j=1}^{n} \mathbf{P_{n-j+1}F_{n-j+1}}\left(L_{n-j+1}-L_{n-j}\right)\right)}_{\mathbf{U}} \begin{bmatrix} \psi(0) \\ N(0) \\ \phi(0) \\ d\phi(0)/dx \\ M(0) \\ V(0) \end{bmatrix} \tag{26}$$

where $\mathbf{U}$, the product of all of the point and field transfer matrices (a result of the semigroup property of $\mathbf{\Phi}$), is called the *system transfer matrix*. This matrix is the state transition matrix from the fixed end to the free end, across all of the beam segments and junctions. As will be demonstrated, this is the matrix that is used in the eigensolution of the structure.

Substituting Eq. (25) into Eq. (26) and examining the 2nd, 5th, and 6th equations of the resulting linear system reveals

$$\begin{bmatrix} 0 \\ 0 \\ 0 \end{bmatrix} = \begin{bmatrix} U_{2,2} & U_{2,5} & U_{2,6} \\ U_{5,2} & U_{5,5} & U_{5,6} \\ U_{6,2} & U_{6,5} & U_{6,6} \end{bmatrix} \begin{bmatrix} N(0) \\ M(0) \\ M(0) \end{bmatrix} \tag{27}$$

where $U_{i,j}$ is the i,j component of the system transfer matrix $\mathbf{U}$. Solving the characteristic equation of the matrix appearing in Eq. (27) yields the natural frequencies ω of the structure, and hence, the conditions for the existence of non-trivial solutions to Eq. (27). The resulting characteristic equation is shown to reduce to the standard eigenvalue formulas for cantilevered beams (with or without tip mass) in (Reissman et al., 2011).

3.2 Mode shapes

To compute the mode shapes, Eq. (4) is again revisited, this time evaluated between the fixed end and an arbitrary point along the structure:

$$\begin{bmatrix} \psi(x) \\ N(x) \\ \phi(x) \\ d\phi(x)/dx \\ M(x) \\ V(x) \end{bmatrix} = \mathbf{\Phi}(x,0) \begin{bmatrix} \psi(0) \\ N(0) \\ \phi(0) \\ d\phi(0)/dx \\ M(0) \\ V(0) \end{bmatrix} \tag{28}$$

The first equation in Eq. (28) is evaluated for the mode shape:

$$\begin{aligned} \phi(x) &= \left[\mathbf{\Phi}(x,0)\right]_{3,2} N(0) + \left[\mathbf{\Phi}(x,0)\right]_{3,5} M(0) + \left[\mathbf{\Phi}(x,0)\right]_{3,6} V(0) \\ &= \left\{ -\left[\mathbf{\Phi}(x,0)\right]_{3,2} (k_1 + k_2\sigma) + \left[\mathbf{\Phi}(x,0)\right]_{3,5} + \left[\mathbf{\Phi}(x,0)\right]_{3,6} \sigma \right\} M(0) \end{aligned} \tag{29}$$

where the constants are computed according to the following conditions:

case $U_{5,2} \neq 0$:

$$k_1 = \frac{U_{5,5}}{U_{5,2}},\ k_2 = \frac{U_{5,6}}{U_{5,2}},\ \text{and}\ \sigma = \frac{U_{6,2}U_{5,5} - U_{6,5}U_{5,2}}{U_{6,6}U_{5,2} - U_{6,2}U_{5,6}} \tag{30a}$$

case $U_{6,2} \neq 0$:

$$k_1 = \frac{U_{6,5}}{U_{6,2}},\ k_2 = \frac{U_{6,6}}{U_{6,2}},\ \text{and}\ \sigma = \frac{U_{6,2}U_{5,5} - U_{6,5}U_{5,2}}{U_{6,6}U_{5,2} - U_{6,2}U_{5,6}} \tag{30b}$$

case $U_{5,2} = 0$ and $U_{6,2} = 0$:

$$k_1 = 0,\ k_2 = 0,\ \text{and}\ \sigma = -\frac{U_{6,5}}{U_{6,6}} \tag{30c}$$

In Eq. (29), the scaling factor $M(0)$ is not retained: instead the mode shapes are scaled in order to satisfy the appropriate orthogonality conditions, as discussed in section 4.2.

4. Solution to electromechanical EOMs via modal analysis

4.1 Calculation of base excitation contribution

In this section, the EOMs are solved using a modal decoupling procedure. However, before this can be accomplished, the external forcing term $f(x,t)$ appearing in Eq. (10) must be evaluated. This term represents an applied transverse force/length along the beam segments. A common use for this term is pressure loads due to flowing media into which the structure is immersed. In the present scenario, this load is the apparent inertial loading due to the excitation of the base in the vertical direction.

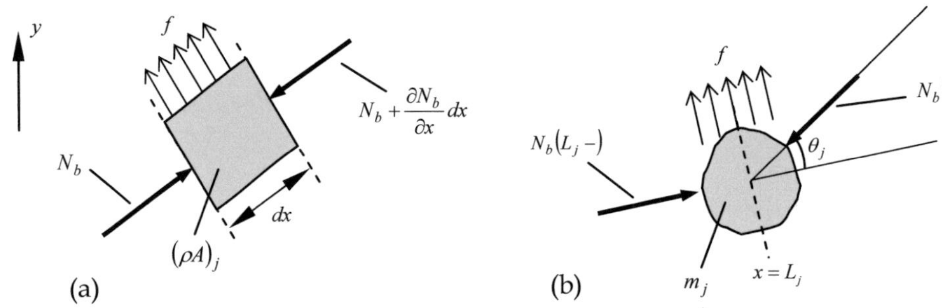

Fig. 4. Forces due to base excitation on a beam element (a) and on a lumped mass located at $x = L_j$ (b).

In Fig. 4, the forces due to the apparent inertial loads from the base excitation are shown for an arbitrary element of a beam segment and a lumped mass. Due to rotations at the lumped mass interfaces, the inertial loads are not strictly transverse or axial, but have components in both directions. The absolute orientation of each component determines how the base excitation affects it; this orientation is the sum of the relative angles between the joints between the base and the component. Only the normal force due to base excitation, denoted N_b, is included here; the other forces and moments have already been accounted for in section 2.2.

A balance of forces in the transverse and axial directions for the element shown in Fig. 4(a) gives

$$f(x,t) = -(\rho A)_j \cos\left(\sum_{i=0}^{j-1} \theta_i\right) \frac{d^2 y(t)}{dt^2} \tag{31}$$

$$\text{and} \quad \frac{\partial N_b(x,t)}{\partial x} = (\rho A)_j \sin\left(\sum_{i=0}^{j-1} \theta_i\right) \frac{d^2 y(t)}{dt^2} \tag{32}$$

respectively, where $\theta_0 \equiv 0$. Eq. (32) can be integrated to get

$$N_b(L_j -) - N_b(L_{j-1} +) = (\rho A)_j l_j \sin\left(\sum_{i=0}^{j-1} \theta_i\right) \frac{d^2 y(t)}{dt^2} \tag{33}$$

Similarly, a balance of forces in the transverse and axial directions for the lumped mass shown in Fig. 4(b) gives

$$f(x,t) = \left[-m_j \cos\left(\sum_{i=0}^{j-1} \theta_i \right) \frac{d^2 y(t)}{dt^2} + N_b(L_j +) \sin\theta_j \right] \delta(x - L_j) \tag{34}$$

$$\text{and } N_b(L_j -) - N_b(L_j +) \cos\theta_j = -m_j \sin\left(\sum_{i=0}^{j-1} \theta_i \right) \frac{d^2 y(t)}{dt^2} \tag{35}$$

respectively. Combining Eqs. (31,34) gives

$$\begin{aligned} f(x,t) = -\frac{d^2 y(t)}{dt^2} \sum_{j=1}^{n} \Bigg(\left\{ (\rho A)_j \left[H(x - L_{j-1}) - H(x - L_j) \right] + m_j \delta(x - L_j) \right\} \cos\left(\sum_{i=0}^{j-1} \theta_i \right) \Bigg) \\ + N_b(L_j +) \sin\theta_j \delta(x - L_j) \end{aligned} \tag{36}$$

where

$$N_b(L_n +) = 0 \text{ and}$$

$$N_b(L_j +) = -\left[(\rho A)_{j+1} l_{j+1} + m_{j+1} \right] \sin\left(\sum_{i=0}^{j} \theta_i \right) \frac{d^2 y(t)}{dt^2} + N_b(L_{j+1} +) \cos\theta_{j+1} \tag{37}$$

which can be evaluated inductively.

4.2 Modal decoupling

The EOMs for a single beam segment have been derived in section 2.2 and subsequently used to develop the field transfer matrix for such a segment. Using the transfer matrix method, the natural frequencies and mode shapes have been calculated. Now, the time response is found by decoupling the partial differential equations into a system of ordinary differential equations, one for each mode. By concatenating Eq. (10-11) for each segment, the following EOMs, which apply over the entire structure, can be found:

$$\begin{aligned} \sum_{j=1}^{n} \Bigg\{ \left[(\rho A)_j \frac{\partial^2 w(x,t)}{\partial t^2} + (EI)_j \frac{\partial^4 w(x,t)}{\partial x^4} \right] \left[H(x - L_{j-1}) - H(x - L_j) \right] + \\ + \vartheta_j \left[\frac{d\delta(x - L_L)}{dx} - \frac{d\delta(x - L_R)}{dx} \right] v(t) \Bigg\} = f(x,t) \end{aligned} \tag{38}$$

$$q(t) = \sum_{j=1}^{n} \vartheta_j \left[\left. \frac{\partial w(x,t)}{\partial x} \right|_{x=L_R} - \left. \frac{\partial w(x,t)}{\partial x} \right|_{x=L_L} \right] - \sum_{j=1}^{n} C_j v(t) \tag{39}$$

where the external forcing due to base excitation can be evaluated using Eq. (36).

To orthonormalize the mode shapes, Eq. (38) is considered when there are no external loads (including electrical), i.e. $v(t) \equiv 0$ and $f(x,t) \equiv 0$. Substituting the modal decomposition given by Eq. (1), and assuming a sinusoidal time response gives

$$\omega_r^2 \sum_{j=1}^{n} \left\{ \left[(\rho A)_j \phi_r(x) \right] \left[H(x - L_{j-1}) - H(x - L_j) \right] \right\} = \sum_{j=1}^{n} \left\{ \left[(EI)_j \frac{d^4 \phi_r(x)}{dx^4} \right] \left[H(x - L_{j-1}) - H(x - L_j) \right] \right\} \quad (40)$$

for each term r in the modal expansion. Subsequently, Eq. (40) is multiplied by $\phi_s(x)$ and integrated from $x = 0$ to $x = L_n$. After integrating by parts and applying the boundary and intermediate conditions (i.e. across the lumped masses), the following orthogonality condition is derived:

$$\begin{aligned} \sum_{j=1}^{n} \Bigg[& \int_{L_{j-1}}^{L_j} (\rho A)_j \phi_r(x) \phi_s(x) dx + m_j \phi_r(L_j -) \phi_s(L_j -) + I_j \left. \frac{d\phi_r(x)}{dx} \right|_{x=L_j -} \left. \frac{d\phi_s(x)}{dx} \right|_{x=L_j -} \\ & + (\rho A)_j l_j \psi_r(L_j -) \psi_s(L_j -) + m_j \psi_r(L_j -) \psi_s(L_j -) \Bigg] = \delta_{rs} \end{aligned} \quad (41)$$

where δ_{rs} is the Kronecker delta. If the mode shapes are scaled appropriately such that Eq. (41) is satisfied, then automatically

$$\begin{aligned} \sum_{j=1}^{n} (EI)_j \Bigg[& \int_{L_{j-1}}^{L_j} \frac{d^4 \phi_r(x)}{dx^4} \phi_s(x) dx - \left. \frac{d^3 \phi_r(x)}{dx^3} \right|_{x=L_j -} \left. \frac{d\phi_s(x)}{dx} \right|_{x=L_j -} \\ & + \left. \frac{d^3 \phi_r(x)}{dx^3} \right|_{x=L_{j-1} +} \left. \frac{d\phi_s(x)}{dx} \right|_{x=L_{j-1} +} + \left. \frac{d^2 \phi_r(x)}{dx^2} \right|_{x=L_j -} \left. \frac{d\phi_s(x)}{dx} \right|_{x=L_j -} \\ & - \left. \frac{d\phi_r(x)}{dx} \right|_{x=L_{j-1} +} \left. \frac{d\phi_s(x)}{dx} \right|_{x=L_{j-1} +} \Bigg] = \omega_r^2 \delta_{rs} \end{aligned} \quad (42)$$

is satisfied, thus decoupling Eq. (38). Subsequently, the natural frequencies and mode shape functions derived from the TMM can be adopted into existing piezoelectric energy harvester models for evaluating continuous and discontinuous structures.

4.3 Frequency response functions

Once the EOMs are decoupled by mode, the frequency response functions (FRFs) of the structure can be obtained in a straightforward manner by substituting the modal expansions given by Eq. (1) into Eq. (38-39) and applying the orthogonality conditions of Eqs. (41-42). The decoupled forms of Eqs. (38-39) are given for the rth mode:

$$\frac{d^2 \eta_r(t)}{dt^2} + 2\zeta_r \omega_r \frac{d\eta_r(t)}{dt} + \omega_r^2 \eta_r(t) + \Theta_r v(t) = -(\rho A \gamma)_r \frac{d^2 y(t)}{dt^2} \quad (43)$$

$$\frac{dv(t)}{dt} + \frac{1}{R_l C_0} v(t) = \frac{1}{C_0} \sum_{r=1}^{\infty} \Theta_r \frac{d\eta_r(t)}{dt} \quad (44)$$

where Eq. (43) represents the mechanical equation, in which the modal short-circuit frequencies ω_r are equal to the natural frequencies derived from the TMM. (Setting $v(t) \equiv 0$, i.e. shorting the terminals of the device, is equivalent to decoupling the electrical from the mechanical dynamics.) The modal electromechanical coupling is given by

$$\Theta_r = \sum_{j=1}^{n} \vartheta_j \left[\left.\frac{d\phi_r(x)}{dx}\right|_{x=L_j} - \left.\frac{d\phi_r(x)}{dx}\right|_{x=L_{j-1}} \right] \tag{45}$$

and the modal influence coefficient of the distributed inertial force along the beam is

$$(\rho A\gamma)_r = \sum_{j=1}^{n} \left\{ \left[\int_{L_{j-1}}^{L_j} (\rho A)_j \phi_r(x)dx + m_j\phi_r(L_j -) \right] \cos\left(\sum_{i=0}^{j-1} \theta_i \right) + N_b(L_j +) \sin\theta_j \phi_r(L_j -) \right\} \tag{46}$$

Note that the modal damping term $2\zeta_r\omega_r(d\eta_r(t)/dt)$ has been added at this point, although the value of the modal damping ratio is usually determined experimentally.

Eq. (44) represents the electrical dynamics equation where the terminals of the device are assumed to be placed across an external resistor R_l. The net clamped, i.e. constant strain, capacitance of the piezoelectric material, appearing in Eq. (44), is defined as

$$C_0 = \sum_{j=1}^{n} C_j \tag{47}$$

which in the parallel bimorph configuration is simply the sum of the capacitances of the beam segments. On the right-hand side of Eq. (44), the same Θ_r appearing in Eq. (43) is used to couple the two modal EOMs. It should be noted that only under mass-normalized conditions, i.e. Eq. (41), are these two coupling coefficients equal.

To evaluate the FRFs of the structure, a harmonic base excitation $y(t) = Ye^{i\omega t}$ is assumed. Given that the eigensolutions are derived from Euler-Bernoulli beam theory (resulting in a linear PDE), and the piezoelectric constitutive equations are also linearized (see Eq. (8)), the resulting motion and voltage output are also harmonic at the base excitation frequency. Thus, the relative transverse motion at a point x from the base is given by

$$w(x,t) = W(x)e^{i\omega t} = \sum_{r=1}^{\infty} \frac{(\rho A\gamma)_r \omega^2 Y - \Theta_r V}{\omega_r^2 - \omega^2 + i2\zeta_r\omega_r\omega} \phi_r(x) e^{i\omega t} \tag{48}$$

where $i = \sqrt{-1}$, and the voltage output is

$$v(t) = Ve^{i\omega t} = \frac{\dfrac{1}{C_0} \displaystyle\sum_{r=1}^{\infty} \frac{i(\rho A\gamma)_r \omega^3 \Theta_r Y}{\omega_r^2 - \omega^2 + i2\zeta_r\omega_r\omega}}{i\omega + \dfrac{1}{R_l C_0} + \dfrac{1}{C_0} \displaystyle\sum_{r=1}^{\infty} \frac{i\omega \Theta_r^{\,2}}{\omega_r^2 - \omega^2 + i2\zeta_r\omega_r\omega}} e^{i\omega t} \tag{49}$$

Finally, the current output of the device can be found from $i(t) = v(t)/R_l$ and the power from $p(t) = [v(t)]^2/R_l$.

5. Case studies

In order to demonstrate the use of the TMM for analysis of multi-segmented beam structures, a few simple examples are given. Fig. 5 depicts the two cases under consideration. Fig. 5(a) shows a bimorph beam with varying piezoelectric layer coverage starting from the base and extending to the point x_{div}, the dividing line between the two segments. Note that the point transfer matrix between the two segments, given in Eq. (24), reduces to the identity matrix in this case. Fig. 5(b) shows a bimorph beam with varying center joint angle. The special case $\theta_1 = 0^\circ$ corresponds to a standard cantilevered bimorph, whereas the case $\theta_1 = 90^\circ$ corresponds to an L-shaped structure, previously studied by (Erturk et al., 2009). In both cases, the overall length of the structure is kept constant. The geometry and material properties of the device are listed in Table 1.

Fig. 6 plots the variation in fundamental (short-circuit) natural frequency as x_{div} is varied between 0 and L, the overall length of the structure, and the layer thickness ratio (over the segment with piezoelectric coverage) is varied between $t_p/t_s = 0.05$ and $t_p/t_s = 3$. These curves indicate that in each case the beam with full coverage has a higher natural frequency than the beam with no coverage, with agrees with the fact that the effective stiffness increases when the piezoelectric layers are added, i.e. $(EI)_1 > (EI)_2$. Furthermore, this effect is exacerbated with larger layer thickness ratio. Somewhat surprisingly, the natural frequency has a maximum at a point of partial coverage; this maximum shifts towards $x_{div} = L$ as t_p/t_s increases. This phenomenon can be explained by considering that the partial coverage makes the beam effectively shorter, since the bare substructure section does not contribute to the stiffness as significantly. Increasing the layer coverage past this maximum effectively lengthens the beam, thus decreasing its natural frequency. The reverse effect occurs as the layer coverage decreases further: the bare substructure region dominates, reducing the natural frequency.

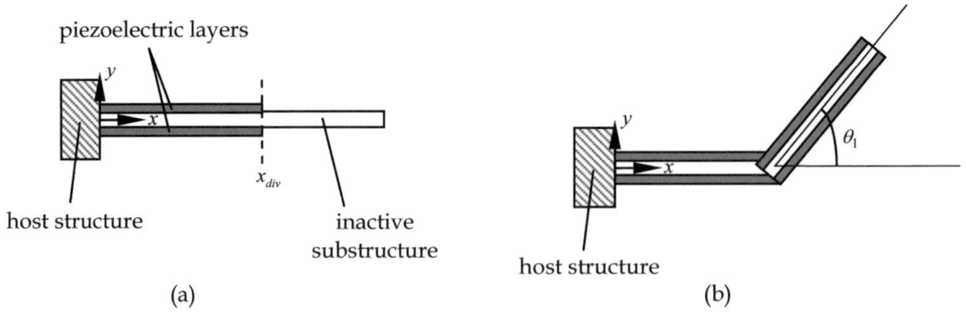

Fig. 5. (a) Layout of bimorph structure with partial piezoelectric layer coverage. (b) Layout of bimorph structure with a variable-angle bend at half its overall length.

Beam properties:		
L	length	100 mm
b	width	20 mm
t_s	thickness of substructure	0.75 mm
t_p	thickness of PZT layer	0.5 mm
ρ_s	density of substructure	8070 kg/m³
ρ_p	density of PZT	7800 kg/m³
$c_{11.s}$	Young's modulus of substructure	102 GPa
c_{11}^E	Young's modulus of PZT	66 GPa
e_{31}	piezoelectric constant	-12.54 C/m²
ε_{33}^S	permittivity	15.93 nF/m

Table 1. Geometry and material properties.

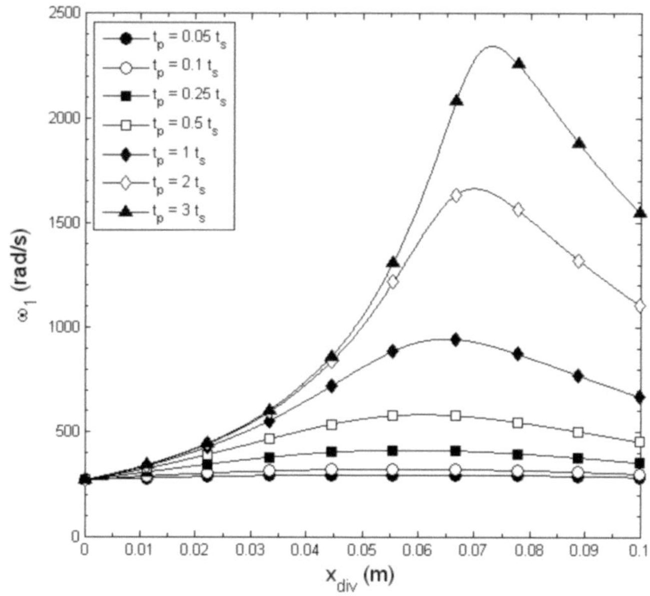

Fig. 6. Variation in fundamental (short-circuit) natural frequency with respect to patch coverage and layer thickness ratio.

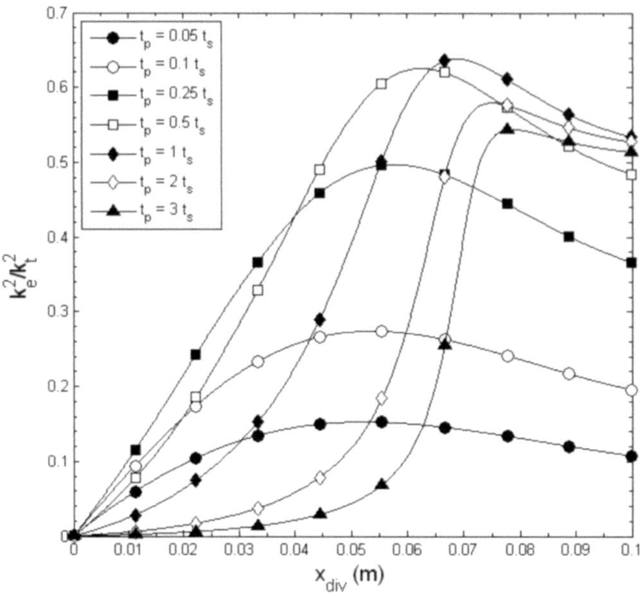

Fig. 7. Variation in coupling coefficient ratio with respect to patch coverage and layer thickness ratio.

To quantify the effects on the transduction capabilities of these two designs, the dimensionless electromechanical coupling coefficient is employed. The device (i.e. entire structure) coupling coefficient for the rth mode is given by $k_e^2 = \Theta_r^2 / \left(C_0 \omega_r^2\right)$, a term that has been used to non-dimensionalize power in the literature, e.g. (Shu and Lien, 2006; Liao and Sodano, 2008; Wickenheiser and Garcia, 2010c). It has been shown by (Wickenheiser, 2011) that this term can be written in the form

$$k_e^2 = k_t^2 \kappa_M \kappa_{EI} \tag{50}$$

for single segment beams (with or without tip mass), where $k_t^2 = e_{31}^2 / \left(c_{11}^E \varepsilon_{33}^S\right)$ is the piezoelectric material coupling coefficient, κ_M is a dimensionless term that depends on the distribution of inertia between the beam and tip mass, and κ_{EI} is a dimensionless term that depends on the effective stiffness of the beam. Although the simple decomposition of Eq. (50) is not proven for more general structures, it can be shown that the ratio k_e^2 / k_t^2 is constant for a specific geometry.

In the present study, the ratio k_e^2 / k_t^2 is plotted, isolating the effects of geometry on the electromechanical coupling of the device. In Fig. 7, the piezoelectric layers are extended out from the base to x_{div}. As the coverage decreases to 0, $k_e^2 / k_t^2 \to 0$ as expected. As it increases, there is a maximum before $x_{div} = L$, similar to Fig. 6. This indicates that the extra coverage at the end of the beam is not utilized efficiently, which is due to the relative lack of strain there. Furthermore, increasing the piezoelectric layer thickness does not result in increased coupling past a certain point; this phenomenon is explored in more detail in (Wickenheiser, 2011). Briefly, the increased thickness increases the stiffness more quickly than the dimensional coupling coefficient, resulting in decreased k_e^2.

Figs. 8-10 show the results of the center joint angle parameter study for the design depicted in Fig. 5(b). In this study, the center joint angle θ_1 is varied between 0 and 180 deg. Both beam segments are identical bimorphs. As with the previous case, increasing the piezoelectric layer thickness increases the bending stiffness and therefore the natural frequencies, as shown in Fig. 8. This plot also indicates that the fundamental frequency increases with center joint angle, which causes an effective shortening of the beam. Fig. 9 shows the variation in the ratio of the first two natural frequencies ω_2/ω_1 . First, these results indicate that this ratio is independent of the thickness ratio of the layers; this result is confirmed by (Wickenheiser, 2011), who shows that this ratio is only a function of the ratio of non-dimensional eigenvalues of the device. Moreover, as the device is folded back on itself, i.e. θ_1 increases, the first natural frequency increases while the second decreases, both converging to a common value. As mentioned by (Erturk et al., 2009), this convergence is useful for broadband energy harvesting from a range of excitation frequencies between these two natural frequencies. Furthermore, energy exchange between modes becomes possible when the natural frequencies are commensurable (Nayfeh and Mook, 1979), a useful mechanism for exciting a higher frequency mode from a lower frequency excitation. Finally, Fig. 10 indicates that the coupling coefficient is relatively insensitive to center join angle. This insensitivity is mainly due to the small impact on the strain at the root, where most of the energy is harvested, by the rotation of the relatively strain-free tip segment.

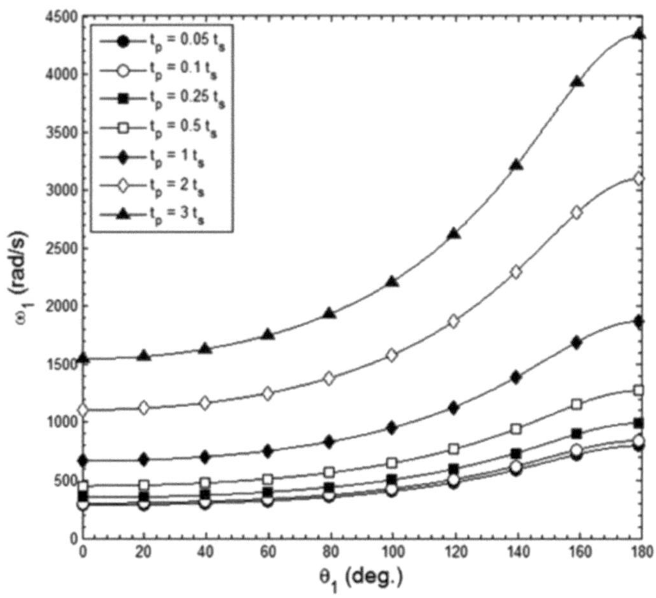

Fig. 8. Variation in fundamental (short-circuit) natural frequency with respect to center joint angle and layer thickness ratio.

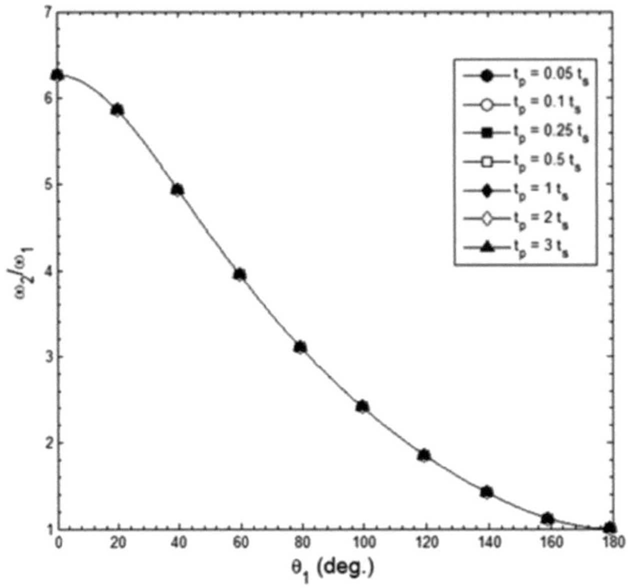

Fig. 9. Variation in ratio of first two natural frequencies with respect to center joint angle and layer thickness ratio.

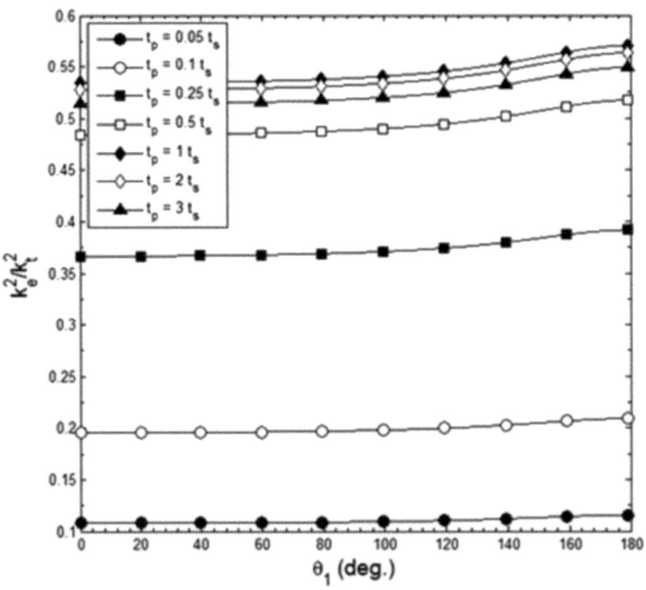

Fig. 10. Variation in coupling coefficient ratio with respect to center joint angle and layer thickness ratio.

6. Conclusions

This chapter presents an electromechanical modeling technique for computing the eigensolution and frequency transfer functions for segmented beams using the transfer matrix method. This method allows for multiple discontinuities in the beam structure, for example partial layer coverage, discontinuities in cross section, angles between members, and multiple lumped masses along the length. The electromechanical equations of motion, including distributed inertial effects from the base excitation, are derived in general terms for various piezoelectric layering configurations. Transfer matrices are derived for each continuous segment and each discontinuity, and a system transfer function is described using the semigroup property of state transition matrices.

This method is general enough to be effective at analyzing many configurations of piezoelectric structures without requiring a re-derivation starting from first principles. Axial and transverse dynamics are shown to be decoupled, leading to block diagonal transfer matrices. Furthermore, it is shown that the size of the eigenvalue problem does not grow with added complexity to the structure, unlike finite element methods. In this chapter, this method is applied to structures with partial piezoelectric layer coverage, taking into account the discontinuity in cross section at the end of the piezoelectric layers. Additionally, variations in center angle are discussed, including the special cases of L-shaped and reflex beams. The natural frequencies and electromechanical coupling coefficients are computed and discussed, displaying the usefulness and versatility of this technique in the design and analysis of complex structures.

7. References

(1987). *IEEE Standard on Piezoelectricity*, IEEE, New York, NY.

Anton, S. R. & Sodano, H. A. (2007). A review of power harvesting using piezoelectric materials (2003–2006). *Smart Materials and Structures*, Vol.6, No.3, (June 2007), pp. R1–R21, ISSN 0964-1726.

Beer, F. P. & Johnson, Jr. E. R. (1992). *Mechanics of Materials* (2nd), McGraw-Hill, ISBN 0-07-004340-X, New York, NY.

Dietl, J. M. & Garcia, E. (2010). Beam shape optimization for power harvesting. *Journal of Intelligent Material Systems and Structures*, Vol.21, pp. 633–646, ISSN 1045389X.

duToit N. E., Wardle, B. L. & Kim, S. (2005). Design Considerations for MEMS-Scale Piezoelectric Mechanical Vibrations Energy Harvesting. *Integrated Ferroelectrics*, Vol.71, pp. 121–160, ISSN 1058-4587.

Erturk, A. & Inman, D. J. (2008). A Distributed Parameter Electromechanical Model for Cantilevered Piezoelectric Energy Harvesters. *Journal of Vibrations and Acoustics*, Vol.130, No.4, 041002, ISSN 10489002.

Erturk, A.; Renno, J. M. & Inman, D.J. (2009). Modeling of piezoelectric energy harvesting from an L-shaped beam-mass structure with an application to UAVs. *Journal of Intelligent Materials Systems and Structures*, Vol.20, pp. 529–544, ISSN 1045-389X.

Guyomar, D.; Badel, A.; Lefeuvre, E. & Richard, C. (2005). Toward energy harvesting using active materials and conversion improvement by nonlinear processing. *IEEE Transactions on Ultrasonics, Ferroelectrics, and Frequency Control*, Vol.52, No.4, (April 2005), pp. 584–595, ISSN 08853010.

Inman, D. J. (2007). *Engineering Vibration* (3rd), Pearson, ISBN 0-13-228173-2, Upper Saddle River, NJ.

Karami, M. A. & Inman, D. J. (2011). Electromechanical Modeling of the Low-Frequency Zigzag Micro-Energy Harvester. *Journal of Intelligent Material Systems and Structures*, Vol.22, No. 3, pp. 271–282, ISSN 1045389X.

Liao, Y. & Sodano, H. A. (2008). Model of a single mode energy harvester and properties for optimal power generation. *Smart Materials and Structures*, Vol.17, No.6, 065026, ISSN 09641726.

Murray, R. & Rastegar, J. (2009). Novel Two-Stage Piezoelectric-Based Ocean Wave Energy Harvesters for Moored or Unmoored Buoys. *Proceedings of SPIE*, ISSN 0277-786X, San Diego, CA, March, 2009.

Nayfeh, A. H. & Mook, D. T. (1979). *Nonlinear Oscillations*, Wiley, ISBN 0471121428, New York, NY.

Pestel, E. & Leckie, F. A. (1963). *Matrix Methods in Elastomechanics*, McGraw-Hill, ASIN B000BLO2YI, New York, NY.

Reissman, T.; Dietl, J. M. & Garcia, E. (2007). Modeling and Experimental Verification of Geometry Effects on Piezoelectric Energy Harvesters. *Proceedings of 3rd Annual Energy Harvesting Workshop*, Santa Fe, NM, February, 2007.

Reissman, T., Wickenheiser, A. M. & Garcia, E. (2011) Closed Form Solutions for the Dynamic Response of Piezoelectric Energy Harvesting Structures with Non-Uniform Geometries. (in preparation).

Roundy, S.; Leland, E. S.; Baker, J.; Carleton, E.; Reilly, E.; Lai, E.; Otis, B.; Rabaey, J.M.; Wright, P.K. & Sundararajan, V. (2005). Improving power output for vibration-based energy scavengers. *Pervasive Computing*, Vol.4, No.1, pp. 28–36, ISSN 1526-1268.

Roundy, S. & Wright, P. K. (2004). A piezoelectric vibration based generator for wireless electronics. *Smart Materials and Structures*, Vol.13, No.5, pp. 1131–1142, ISSN 09641726.

Roundy, S.; Wright, P. K. & Rabaey, J. (2003). A study of low level vibrations as a power source for wireless sensor nodes. *Computer Communications*, Vol.26, No.11, pp.1131–1144, ISSN 0140-3664.

Sodano, H. A., Park, G. & Inman, D. J. (2004). Estimation of Electric Charge Output for Piezoelectric Energy Harvesting. *Strain*, Vol.40, No.2, pp. 49–58, ISSN 00392103.

Shu, Y. C. & Lien, I. C. (2006). Analysis of power output for piezoelectric energy harvesting systems. *Smart Materials and Structures*, Vol.15, No.6, pp. 1499–1512, ISSN 0964-1726.

Tieck, R. M.; Carman, G. P. & Lee, D. G. E. (2006). Electrical Energy Harvesting Using a Mechanical Rectification Approach. *Proceedings of IMECE*, pp. 547–553, ISBN 0791837904.

Wickenheiser, A. M. (2011). Design Optimization of Linear and Nonlinear Cantilevered Energy Harvesters for Broadband Vibrations. *Journal of Intelligent Material Systems and Structures*, Vol. 22, No. 11, pp. 1213-1225, ISSN 1045389X.

Wickenheiser, A. & Garcia, E. (2010a). Design of energy harvesting systems for harnessing vibrational motion from human and vehicular motion. *Proceedings of SPIE*, ISSN 0277-786X, San Diego, CA, March, 2010.

Wickenheiser, A. M. & Garcia, E., (2010b). Broadband vibration-based energy harvesting improvement through frequency up-conversion by magnetic excitation. *Smart Materials and Structures*, Vol.19, No.6, 065020, ISSN 09641726.

Wickenheiser, A. M. & Garcia, E. (2010c). Power Optimization of Vibration Energy Harvesters Utilizing Passive and Active Circuits. *Journal of Intelligent Materials Systems and Structures*, Vol.21, No.13, pp. 1343–1361, ISSN 1045389X.

Wu, W. J., Wickenheiser, A. M., Reissman, T. & Garcia, E. (2009). Modeling and experimental verification of synchronized discharging techniques for boosting power harvesting from piezoelectric transducers. *Smart Materials and Structures*, Vol.18, No.5, 055012, ISSN 0964-1726.

4

Bandwidth Enhancement: Correcting Magnitude and Phase Distortion in Wideband Piezoelectric Transducer Systems

Said Assous[1], John Rees[3], Mike Lovell[1], Laurie Linnett[2] and David Gunn[3]
[1]*Ultrasound Research Laboratory, University of Leicester*
[2]*Fortkey Ltd*
[3]*Ultrasound Research Laboratory, British Geological Survey*
United Kingdom

1. Introduction

Acoustic ultrasonic measurements are widespread and commonly use transducers exhibiting resonant behaviour due to the piezoelectric nature of their active elements, being designed to give maximum sensitivity in the bandwidth of interest. We present a characterisation of such transducers that provides both magnitude and phase information describing the way in which the receiver responds to a surface displacement over its frequency range. Consequently, these devices work efficiently and linearly over only a very narrow band of their overall frequency range. In turn, this causes phase and magnitude distortion of linear signals. To correct for this distortion, we introduce a software technique, which considers only the input and the final output signals of the whole system which is therefore generally applicable to any acoustic system. By correcting for the distortion of the magnitude and phase responses, we have ensured the signal seen at the receiver replicates the desired signal. We demonstrate a bandwidth extension on the received signal from 60-130 kHz at -6dB to 40-200 kHz at -1dB in a test system. The linear chirp signal we used to demonstrate this method showed the received signal to be almost identical to the desired linear chirp. Such system characterisation will improve ultrasonic techniques when investigating material properties by maximising the accuracy of magnitude and phase estimations.

Piezoelectric transducers are used both as transmitters and receivers in many ultrasonic applications, including non-destructive testing, underwater sonar, radar and medical imaging (Blitz & Simpson, 1996; Urick, 1983; Greenleaf, 2001; Rihaczek, 1969). The transducer outputs are, however, significantly affected by the coupling between the transducer and the other components (e.g. the amplifier, and medium in which the energy propagates) (Fano, 1950) and the bandwidth of the system is also limited by the electro-acoustic performance of the transducers. A number of hardware techniques have been developed towards achieving a flat, broadband frequency response, using matching networks (Schmerr, 2006; Youla, 1964). The load is usually modeled as a resistor and capacitor (Reeder, 1972) or as a simple four-element circuit (Anderson, 1979). The problem with this approach is that, typically, the frequency responses of piezoelectric elements have resonant characteristics, which are difficult to accurately model using the passive component matching networks normally used. In most cases, improved results can be obtained if the network suggested by one of the techniques listed in (Schmerr, 2006; Youla, 1964; Reeder, 1972; Anderson, 1979) is used as a starting point for the hardware optimization approach, which in turn accounts for frequency dependent radiation in the equivalent circuits of the matching networks.

Recently, Doust (Doust, 2001) and Dix (Dou, 2000) introduced a hardware technique, in which they demonstrate improved overall phase linearity, efficiency and amplitude response of transfer functions, in an electro-acoustic system. Doust and Dix (Doust, 2001) sought to improve the accuracy of wave shape measurements and transducer response through compensating their system, comprising amplifiers, filters, and analog-to-digital converters. This was achieved by adding electronic circuits between the amplifier and transducer, and removing phase and amplitude distortion over a frequency spectrum, through a technique they refer to as equalisation. Distortion of the output signal in ultrasonic systems may be caused by many factors within any of the elements of the whole system, not only the transducer elements alone. Often the physical value (e.g. pressure) and the distorted waveform resulting from the conversion processes are repetitive with respect to time, for example current waveforms in alternating current power systems. At any frequency, the transducer, amplifier, filter, or A/D converter can distort the signals introducing errors in the amplitude, phase or both, which can in turn introduce distortion in reconstructed waveforms. The hardware equalisation of Doust (Dou, 2000) achieves this by taking into account all the subsystems and equalising each subsystem in turn.

This hardware method requires a knowledge of the transfer functions of all the components in the system, which maybe difficult to determine (e.g. transfer function of the medium). On the other hand, the software approach, described below, achieves the same but in a single operation, without knowledge of the transfer functions. Consequently, it has the potential to compensate amplitude and phase distortions in whole systems. In essence, we consider the overall system as a 'black-box' and attempt to correct the output by compensating the input on the basis of the system phase and magnitude frequency-responses.

2. Methods

This chapter describes a software method and associated procedures for characterising a system in terms of its magnitude and phase response with respect to frequency; this is then applied experimentally to improve the effective bandwidth of the whole system. The need for transducer characterisation is highlighted through the realisation that the performance of a system may not be known and that assumptions are often made regarding the signal being injected into the medium. Our strategy is to ensure signals with precisely known amplitude and phase are used as inputs to subsequent signal processing methods. Tone burst signals, used for the sensor characterisation, were produced using a piezoelectric transducer driven by an Agilent 33120A function generator. To demonstrate this methodology as a means of improving bandwidth, a series of measurements were performed using ultrasound transducers developed by Alba Ultrasound Ltd. These underwater transducers were designed to have a wide bandwidth with a centre frequency between 100-130 kHz, operating effectively as both transmitters and receivers of ultrasound with 92 mm size diameter and a beam angle of 10° on the primary lobe. The transmitter was driven directly with a 10 V peak-to-peak, 10-cycle tone burst made over a range of frequencies from 40 to 200 kHz sampled at 10MHz and the receiver was connected directly to the oscilloscope. A total of 10000 points were recorded for each waveform at each measurement frequency. A transmitter-receiver separation of approximately 0.5 m was selected as shown in figure 1. We decided to take 2500 samples for each tone-burst to ensure 4 kHz resolution. With this window length and frequency resolution, two sets of 41 signals each starting at 40 kHz, and rising in steps of 4 kHz, to 200 kHz were generated. Using frequencies at this step interval enables the frequency

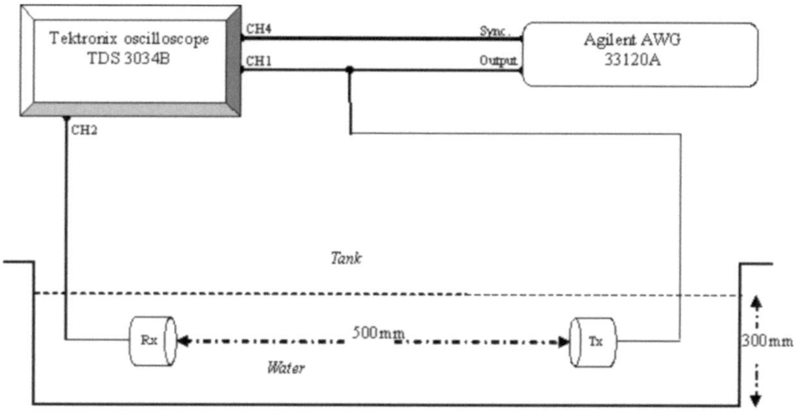

Fig. 1. Schematic diagram of experimental setup.

to be determined exactly on a DFT frequency bin and hence, give an accurate measure and minimise spectrum leakage of the response at that frequency. Furthermore, the length of signals chosen was short enough to avoid interference from multi-path reflections from tank walls. Both sine and cosine signals were generated at each of the corresponding frequencies. These were designated $r0.txt$ and $i0.txt$ for the real (cosine) and imaginary (sine) signals for the first frequency set0, for example. The sets were sequentially numbered from 0 to 40. Examples of the received signals obtained at 48 kHz and the 120 kHz are shown in figure 2a and 2b, respectively. In order to provide a more accurate estimation of the spectrum, we excluded samples affected by the "switch on" and "switch off" of the transducer (see figure 2), taking 700 samples either side of the centre of this 2500 samples long received 'tone-burst' signal for our analyses. This provides 1400 samples about the centre of the 'tone burst' time window avoiding the effects of ringing and reflections. Consequently, we designed two sets of 41 test signals (0...40) each, real and imaginary, in steps of 4 kHz starting at 40 kHz and ending at 200 kHz. Each test signal was transmitted as a continuous sine and cosine wave 'tone-burst' having a $250\mu s$ duration. The discrete Fourier Transform (DFT) of the centre portion (1400 samples) of each received signal was performed.

3. Transducer characterisation and bandwidth enhancement

Having obtained this set of 41 values (via DFT of the sine and cosine sets) over the frequency range, we calculated the response (41 frequency bins) of the system in terms of magnitude and phase with respect to the frequency, as shown in figures 3a and 3b, respectively. We can see in figure 3a, there is a 35 dB variation in magnitude about the centre frequency of 100 kHz. Similarly, the phase in figure 3b is changing rapidly in the centre band of frequencies.

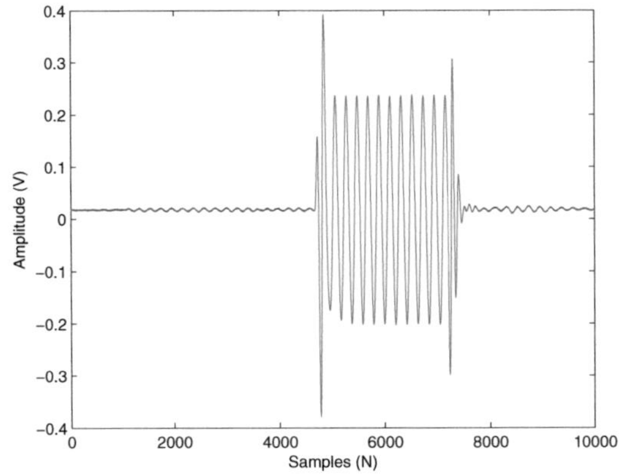

(a) Received signals in the lower transducer sensitivity (48 kHz).

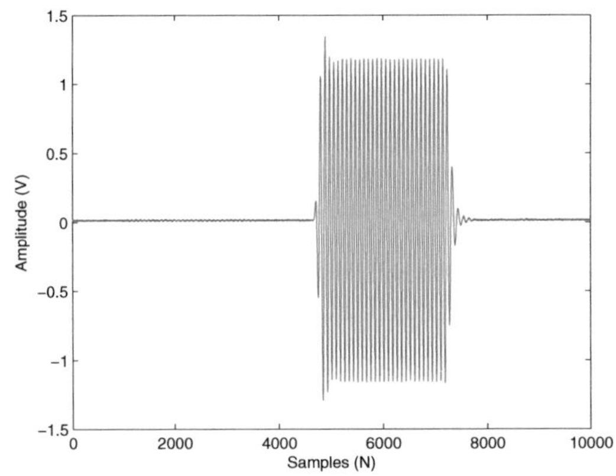

(b) Received signals in the higher transducer sensitivity (120 kHz).

Fig. 2. Examples of received signals.

Using the magnitude and phase responses, described above, the signals were compensated in the time domain in a way similar to inverse filtering, as described below. As a check, the received signals were then used to obtain new sets of magnitude and phase responses as shown in figures 4a and 4b, respectively. Variations in the magnitude and phase responses can be seen to be drastically reduced, following our compensation based on characterising the whole system, magnitude and phase variations being reduced from 35 to 0.6 dB and from 90 to 6 degrees, respectively, resulting in an improvement in the effective bandwidth from 60-130 kHz at -6 dB to 40-200 kHz at -1 dB.

4. Linear chirp compensation

To test the compensated system performance, a linear chirp signal was used. As the characterisation used 41 discrete points in the frequency band, it was necessary to interpolate

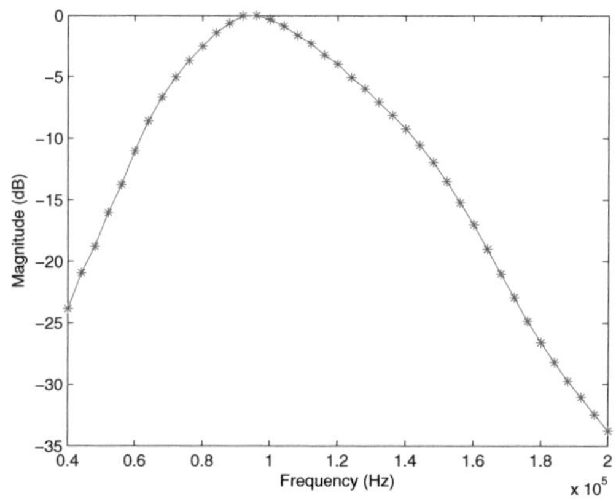

(a) Plot of the 41 magnitude responses related to the 41 transmitted frequencies (35dB variation).

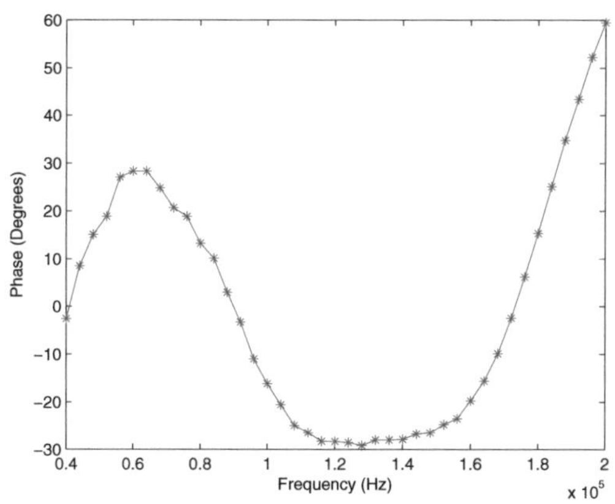

(b) Plot of the 41 phase responses related to the 41 transmitted frequencies (90 degree variation).

Fig. 3. Magnitude and phase responses.

the magnitude and phase responses between these discrete points at all the desired frequencies; we used the *Matlab* 'interp1' function with 'cubic' interpolation. For the purposes of the calibration, 2500 points were used to generate the chirp signal at a sampling frequency of 10 MHz (as described above). The 41 points of the magnitude and phase responses were interpolated to 2500 values as follows:

$$newA = interp1(t, TransferHR(2,:)', new_t, 'cubic'). \tag{1}$$

where *newA* is the amplitude at the required new points, new_t is the time of each of the 2500 new samples points, *t* is the time at the original 41 points and *TransferHR(2,:)* contains the original '41 value' magnitude response. A similar calculation was performed for phase using

$$newP = interp1(t, TransferHR(3,:)', new_t, 'cubic'). \tag{2}$$

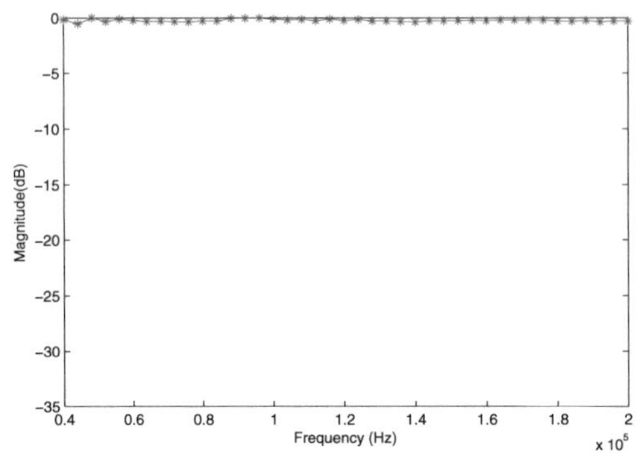

(a) Plot of the 41 magnitude responses related to the compensated transmitted 41 frequencies (0.6 dB variation).

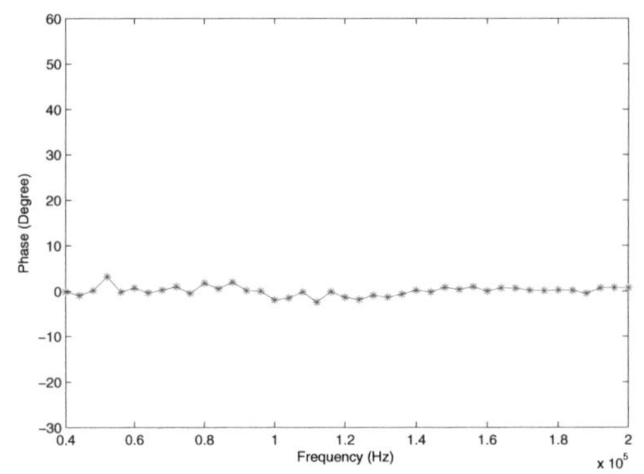

(b) Plot of the 41 phase responses related to the compensated transmitted 41 frequencies (6 degrees variation).

Fig. 4. Magnitude and phase responses after compensation.

where *newP* was an array of 2500 phase values, and *TransferHR(3,:)* contains the original '41values' phase response. Consequently, the compensation for both magnitude and phase was achieved in a single operation in which, the amplitude of the signal will be multiplied by the factor "$max(newA)/newA$" and the phase by "$-newP$". To validate the method, we selected a broadband chirp signal having a frequency range comparable to the transducer response. A Gaussian window was applied to the transmitted signal to minimise the unwanted 'turn on', 'turn off' signals, seen originally in figure 2. These signals are shown in figures 5a and 5b after applying a digital low pass Butterworth filter (0-400 kHz) to eliminate undesirable high frequencies. Our results show a significant difference in the signal excitation (figure 5b) to the one transmitted before (figure 5a). The signal in figure 5b was compensated in relation to the magnitude and phase response 'transferHR' developed as a result of the software characterisation. Figure 6a shows the signal received when the original Gaussian chirp (figure 5a) is transmitted. When the compensated signal was applied to the transmitter (figure 5b), a Gaussian chirp signal was received (figure 6b) similar to the

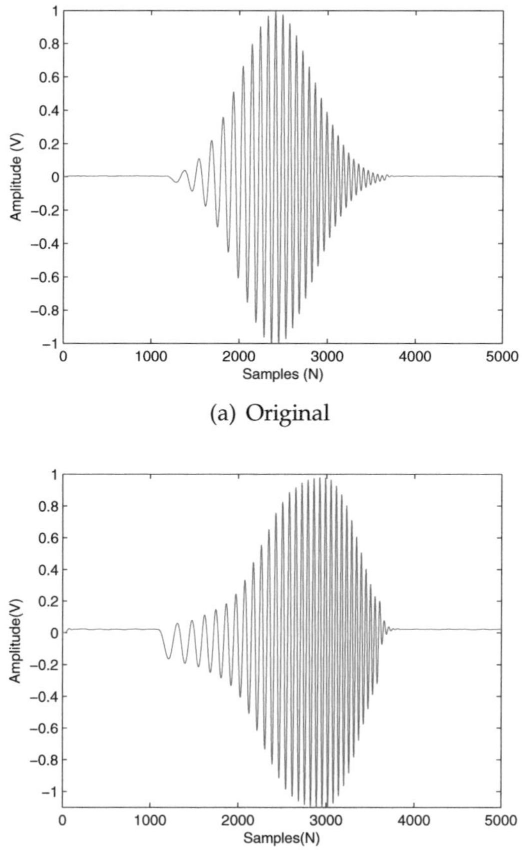

(a) Original

(b) Compensated

Fig. 5. Transmitted gaussian chirps signals [40-200 kHz].

original 'Gaussian' chirp (figure 5a). Thus, the compensation technique can be seen to be effective. An amplitude comparison was undertaken using the envelope of the signals. The envelopes were determined by transforming a time signal into an analytic signal using the Hilbert transform and determining the absolute value of the analytic signal. In figure 7a, the solid curve represents the Hilbert Transform (HT) of the original transmitted signal and the dashed curve represents the HT of the corresponding received signal. Figure 7b shows the HT of the transmitted original signal (solid curve) and the HT of the signal received, following compensated transmission (dashed curve), showing the signals to be almost identical.

5. Results and discussion

In this chapter, we demonstrate a novel software method to improve whole ultrasonic transmitting-receiving systems. Distorting the input signal, on the basis of characterising the magnitude and phase response of the whole system, enabled us to acquire desired signals at the output with little distortion, using piezoelectric transducers in a broadband transmitting and receiving system. Using a linear chirp as a test signal, we validated our method over a

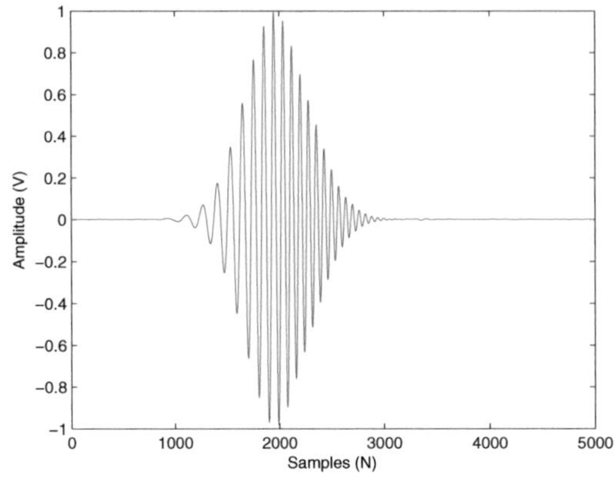

(a) Original

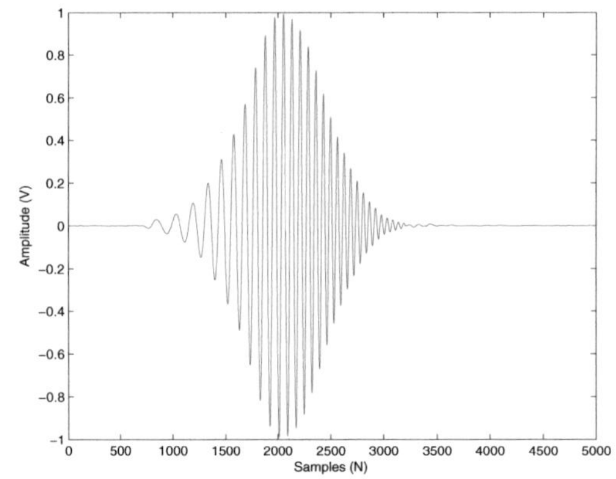

(b) Compensated

Fig. 6. Received gaussian chirps signals [40-200 kHz].

range of frequencies. The results showed close resemblance between the desired and received signals. Our characterisation approach has enabled the effective bandwidth of the system, as a whole, to be significantly improved from 60-130 kHz at -6dB to 40-200 kHz at -1dB. Additionally, such system characterisation is necessary when using ultrasonic techniques to investigate material properties; it is necessary to control signal properties, otherwise the signals will not be sensitive enough to the analysis necessary to identify changes in material properties in terms of changes in their magnitude and phase, for example. Such signals are intended for use in experiments leading to techniques for improved imaging, physical properties characterisation of materials and investigation of material heterogeneity.

The presented technique characterises the effect of the transmission and reception process of acoustic transducers. This enables further measurements to be corrected to remove the effects of the transducers and improve analysis of the wave propagation characteristics.

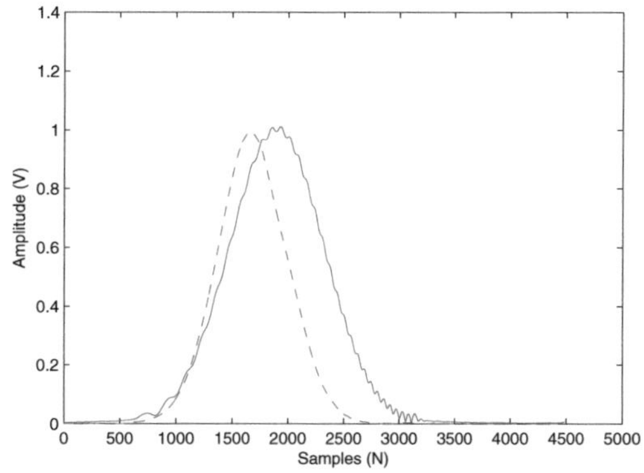

(a) The Hilbert Transforms of the original transmitted signal (solid curve) and original received signal (dashed curve).

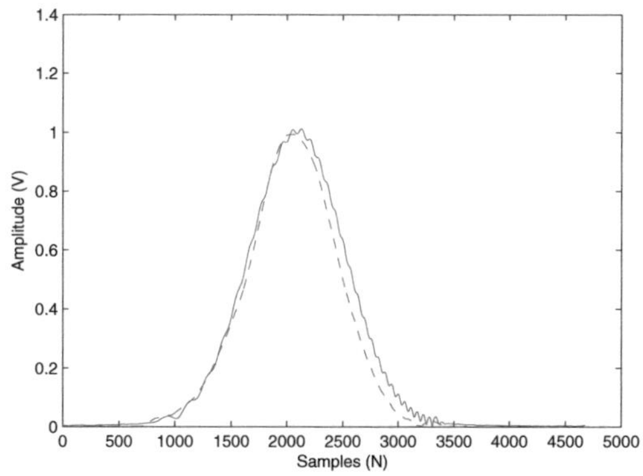

(b) The Hilbert Tranform of the original transmitted signal (solid curve) and the signal received following compensated transmission (dashed curve).

Fig. 7. Compensating the transmitted signal results in the receive signal being almost identical to that originally transmitted (i.e. the desired signal).

6. Acknowledgments

This work was undertaken in the Ultrasound Research Laboratory of the British Geological Survey as part of the Biologically Inspired Acoustic Systems (BIAS) project that is funded by the RCUK via the Basic Technology Programme grant reference number EP/C523776/1. The BIAS project involves collaboration between the British Geological Survey, Leicester University, Fortkey Ltd., Southampton University, Leeds University, Edinburgh University and Strathclyde University.

The work of David Robertson and Victor Murray of Alba Ultrasound Ltd. in the design of the wideband piezo-composite transducers is gratefully acknowledged.

7. References

Blitz, J. & Simpson, G. (1996). Ultrasonic methods of non-destructive testing, Chapman and Hall, (Ed.), New York.

Urick, R. J.(1983). Principles of Underwater Sound, McGraw-Hill, (Ed.), New York.

Rihaczek, A. W.(1969). Principles of high resolution radar, McGraw-Hill, (Ed.), New York.

Greenleaf, J. F (2001). Acoustical medical imaging instrumentation, In: *Encyclopedia of Acoustics*, Crocker, M. J., (Ed.), volume 4, John Wiley and Sons, New York.

Fano, R. M. (1950). Theoretical Limitations of The Broadband Matching of Arbitrary Impedances. *Franklin Institute*, Vol. 244, page numbers (57-83).

Schmerr, L.W.; Lopez-Sanchez, A & Huang, R.(2006) Complete Ultrasonic transducer characterization and use for models and measurements. *Utrasonics*, Vol. 44, page numbers (753-757).

Youla, D.C. (1964) A new theory of broadband matching. *IEEE Trans. Cir. Theory*, Vol. 11, page numbers (30).

Reeder, T.M. Schreve, W.R & Adams, P.L. (1972)A New Broadband Coupling Network for Interdigital Surface wave Transducers. *IEEE Trans. Sonics and Ultrasonics*, Vol. 19, page numbers (466-469).

Anderson, J. & Wilkins, L.(1979) The design of optimum Lumped Broadband Equalizers for ultrasonics Transducers. *J. Acous. Soc*, Vol. 66, page numbers (629).

Doust, P.E. & Dix, J.F.(2001)The impact of improved transducer matching and equalisation techniques on the accuracy and validity of underwater acoustic measurements. In: *Acoustical Oceanography*, Proceeding of Institute of acoustics, Editor: T.G Leighton, G.J Heald, H. Griffiths and G. Griffiths, volume 23 Part2, pages 100-109.

Doust, P.E. (2000) Equalising Transfer Functions for Linear Electro-Acoustic Systems *UK Patent Application*, number 0010820.9.

Rihaczek, A. W.(1969). Principles of high resolution radar, McGraw-Hill, (Ed.), New York, pages (15-20).

Part 2

Applications of Piezoelectric Transducers in Structural Health Monitoring

5

Application of Piezoelectric Transducers in Structural Health Monitoring Techniques

Najib Abou Leyla[1], Emmanuel Moulin[1], Jamal Assaad[1],
Farouk Benmeddour[1], Sébastien Grondel[1] and Youssef Zaatar[2]
[1]Department of OAE, IEMN, UMR CNRS 8520, Université de Valenciennes et du Hainaut Cambrésis, Le Mont Houy,
[2]Applied Physics Laboratory – Lebansese University – Campus,
[1]France
[2]Lebanon

1. Introduction

The technological advances of recent years have contributed greatly to the prosperity of the society. An important element of this prosperity is based on networks of inland, sea, and air transports. However, security in all transport networks remains a major challenge. More specifically, many researches in the field of aeronautics were done to increase the reliability of aircrafts. The themes of NDT (Non Destructive Testing), and more precisely the concept of SHM (Structural Health Monitoring), have thus emerged.

The SHM is a technical inspection to monitor the integrity of mechanical structures in a continuous and autonomous way during its use. Sensors used in this technique being fixed and/or integrated to the structure, it differs from traditional NDT using mobile probes. The first issue is obviously security; a second important issue is reducing financial costs of maintenance. Thus, a new technique that increases reliability and decreases costs of maintenance at the same time seems to be a technological revolution.

Indeed, the traditional inspection methods require planned interventions, and periodic detention of the aircraft, and in some cases the dismantling of some parts. This entire procedure is necessary despite the high costs incurred. Added to that financial aspect, the risk of the occurrence of an unscheduled technical problem between two scheduled inspections is possible. Such a scenario may lead to the replacement of some parts often costly, and fortunately in less frequent cases to air disasters.

2. Background work and contributions of the team

Security in aeronautics being a major issue, regular inspections is needed for maintenance. In fact, these materials are subjected to harsh conditions of operation that may damage them, and thus affect security. Nowadays, traditional inspections induce long immobilization of the aircraft and therefore high costs.

A research project aimed at developing a SHM system based on guided elastic waves and applicable to aeronautic structures has been elaborated by our laboratory. We sought to understand the propagation of ultrasonic waves in the structures, their interaction with damage, and the behavior of different type of sensors. This global vision aims at increasing the reliability of the inspections while decreasing maintenance costs. For this purpose, embedding the transducers to the structure seems to be interesting. Indeed, ultrasonic waves present in the material, called Lamb waves, spread over long distances and interact with damages present in the structure. Their damage detection capabilities has been known for a long time [1,2]. By implanting small piezoelectric transducers into the structures, Lamb waves can be emitted and received and it is theoretically possible to monitor a whole given area [3,4].

Over the last 15 years, the team has therefore studied the different axes of the SHM:

- Characterization of wave propagation in aeronautic materials
- Interaction of ultrasonic waves with damage
- The use of different type of sensors
- Behavior modeling of transducers
- Development of adapted processing tools

The expected benefits of SHM system can thus be summarized in few points:

- Optimization of maintenance plans (decrease immobilisation periods and therefore maintenance costs)
- Increase security (more frequent inspections in an almost real-time)
- Increase the life duration of an aircraft

The theme of SHM started in our team for fifteen years with the researches of Blanquet [5] and Demol [6]. These works were devoted respectively to the study of the propagation of the Lamb waves and the development of multi-elements transducers to generate and receive this type of waves in a material. In the continuity of these researches, E. Moulin has studied the Lamb wave generation [7, 8, 9] and has developed a technique [10] allowing the prediction of the Lamb wave field excited in an isotropic plate by a transducer of finite dimensions. In another work [11], a simple and efficient way of modeling a full Lamb wave emission and reception system was developed. Other works have treated the question of Lamb wave emission and/or reception using thin, surface-bonded PZT transducers. Giurgiutiu [12] has used a "pin force" model to account for the mechanical excitation provided by the emitter. Then, the response of the plate in terms of Lamb waves excitation has been derived analytically in a 2D, plane strain situation. Nieuwenhuis et al [13] have obtained more accurate results by using 2D finite element modeling, which allows a better representation of the transducer behavior. Lanza di Scalea et al [14] have focused on the reception of unidirectionally propagating, straight-crested Lamb or Rayleigh waves by a rectangular transducer.

S. Grondel [15] has optimized the SHM using Lamb waves for aeronautic structures by studying the adapted transducer to generate Lamb waves. Paget [16] has elaborated a technique to detect damage in composite materials. In another work, Duquenne [17] has implemented a hybrid method to generate and receive Lamb waves using a glued-surface transducer in a transient regime. F. El Youbi [18] has developed a sophisticated signal processing technique based on the frequency-time technique and the Fourier transform spatiotemporal to study separately the sensibility of each Lamb mode to damage. More recently F. Benmeddour [19, 20] has studied the interaction of Lamb waves with different type of damage by modelling the diffraction phenomena, and M. Baouahi [21] has considered the 3-D aspect of the Lamb waves generation.

A wide range of work has already been reported on the interaction of Lamb waves with damage and discontinuities such as holes [22, 23], delaminations [24, 25], vertical cracks [26], inclined cracks [27], surface defects [28], joints [29, 30] and thickness variations [31]. More specifically, studies of the interaction with notches have also been carried out. For example, Alleyne et al. [32, 33], have studied the sensitivity of the A0, S0 and A1 modes to notches. In this analysis, they have shown the interest of using the two-dimensional Fourier transform to quantify each scattered Lamb mode. More recently, Lowe et al. [34, 35], have analyzed the behavior of the A0 mode with notches as a function of the width and the depth. The authors [36] have then extended this study to the S0 mode.

Finally, E. Moulin [37] studied and validated experimentally the potential of developing a passive SHM system (without the need on an active source), based on the exploitation of the natural ambient acoustic sources present in an aeronautic structure during flight. This technique has been widely exploited in seismology [38, 39, 40], underwater acoustics [41, 42] and recently ultrasonic [43, 44, 45].

These works contributed to the progress of resolving the different problems related to SHM. They can be synthesized in four main points:

- The modeling of an active complete SHM system (transmission / propagation / reception), in the absence of damage: Development of an efficient modeling tool that takes into account the actual characteristics, including 3-D aspect
- The modeling and interpretation of diffraction phenomena (transmission, reflection, mode conversions) on a number of defect types
- Development of a sophisticated signal processing technique (time-frequency technique and spatiotemporal Fourier transform) to minimize the risk of misinterpretation
- Demonstration of the feasibility of a passive SHM system based on the exploitation of natural ambient acoustic fields.

All of this work has enabled the team to acquire the skills needed to develop a SHM system. The following section is devoted to the presentation of some results for each of the above points.

3. Modeling and results

3.1 Modeling of emission – reception transient system

The work described in this section is intended to present a simple and efficient way of modeling a full Lamb-wave emission and reception system. The advantage of this modular approach is that realistic configurations can be simulated without performing cumbersome modeling and time-consuming computations. Good agreement is obtained between predicted and measured signals.

It will be assumed here that the bonded piezoelectric receiver has a small influence on the wave propagation [46]. As a consequence, the displacement imposed at the plate-receiver interface will be considered to be the surface displacement field associated to the incident Lamb waves. The validity of this assumption is realistic if either the transducer is very thin compared to the plate thickness or very small compared to the shortest wavelength [11].

The emission process is modeled using the finite element - normal mode technique described above. It allows predicting the modal amplitudes of each Lamb mode. Then, each Lamb mode is treated separately as a displacement input on the lower surface of the receiving transducer. In summary, the input of the model is the electrical signal (in volts) applied to the emitter and the output is the electrical signal (in volts) received at the receiver. Therefore, the theoretical and measured results can be directly compared to each other, without any adjustment parameters.

Two experimental configurations have been tested. In each case, two parallelepiped-shaped piezoelectric transducers have been glued on the upper surface of a 6-mm thick aluminum plate. One of them is 3-mm wide, 500-μm thick and 2-cm long and is used as the emitter. In the case which will be referred to as setup #1, the receiver is 3-mm wide, 200-μm thick, 2-cm long and has been placed 20 cm away from the emitter. In setup #2, the emitter-receiver distance is 15 cm and the receiver is 0.5-mm wide, 1-mm thick and 2-cm long (figure 1). This choice for the dimensions of the receivers has been guided by the assumption discussed above.

The electric excitation signal is provided by a standard waveform generator with output impedance 50 Ω. The signal received at the second transducer is directly measured, without any amplification or filtering, using a digital oscilloscope with sampling rate 25 Ms/s and input impedance 1 MΩ. The lateral dimensions of the plate have been chosen large enough as for avoiding parasitic reflections mixed with the useful parts of the received signals. The frequency range considered for the study spreads from 100 to 500 kHz.

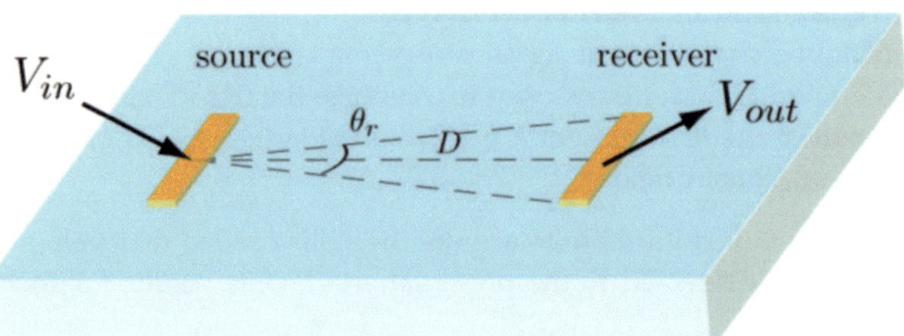

Fig. 1. Source – receiver configuration. Three-dimensional, realistic situation

The first example, presented in figure 2, corresponds to setup #1 where the emitter is excited with an input signal V*in* which is a 5-cycle, 10 V amplitude, Hanning-windowed sinusoid signal with central frequency 450 kHz. The received signal is an electric voltage V*out*. In this case, the first three Lamb modes A0, S0 and A1 are expected.

The measured and predicted waveforms (Fig. 2-a and -b, respectively) appear to be in good agreement. The three main wave packets, corresponding to the three generated Lamb modes, are clearly visible.

The second example corresponds to the results obtained for setup #2, with excitation frequency 175 kHz (Fig. 3). Comments very similar as above can be made. Here again, both the absolute amplitudes as well as the global waveforms are correctly predicted.

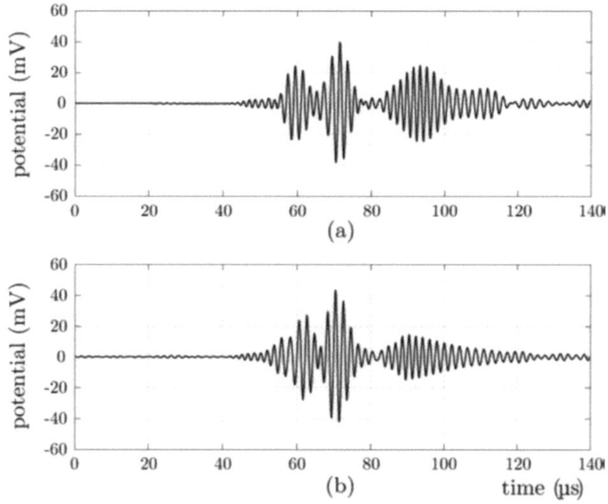

Fig. 2. Electrical potential at the 3-mm wide, 0.2-mm thick receiver (setup #1), for a 450 kHz excitation.(a) Experimental. (b) Predicted.

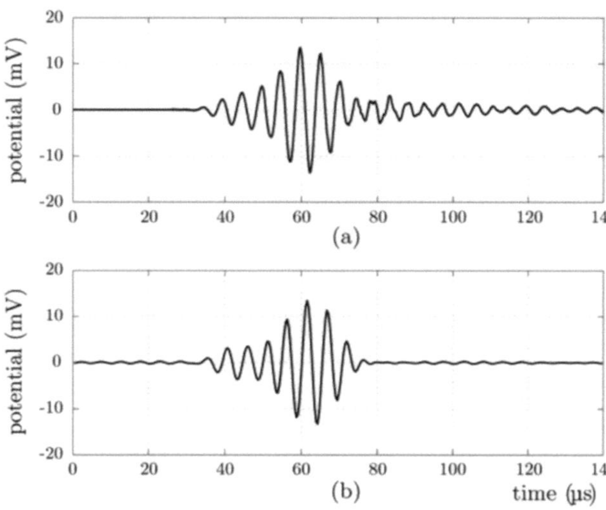

Fig. 3. Electrical potential at the 0.5-mm wide, 1-mm thick receiver (setup #2), for a 175 kHz excitation. (a) Experimental. (b) Predicted.

3.2 Fundamental Lamb modes interaction with symmetrical and asymmetrical discontinuities

The aim of the work presented in this section is to predict the propagation of the fundamental Lamb modes in a structure containing symmetrical [19] and asymmetrical [20] discontinuities in a simple and a fast way. The key point is to decompose a given damage into two elementary types; the symmetrical damage with respect to the median plane and the asymmetrical one in the plate section. The power reflection and transmission coefficients are computed, using two techniques, the finite element method with the help of the Atila code, and the average power flow equation. Indeed, the characterization of damage is done from the average flow of power. It allows determining the power reflection and transmission coefficients.

To calculate these coefficients, normal and tangential displacements, as well as the stresses throughout the plate thickness are needed. It is experimentally impossible. While the developed method, based on the decomposition of normal modes, allows the calculation of the average power flow from only a normal displacement or a tangential one acquired at the surface of the plate. The advantages of this technique are that the time processing of the data is reduced, and it allows a direct comparison with the experimental measurements.

To validate the numerical results, the following experimental study was carried out. The instrumentation used in the experimental investigation is shown schematically in the fig. 4. A pulse, from the pulse generator (HM 8035), is used to simultaneously trigger the oscilloscope (Le Croy type LT344) and the two arbitrary function generators (HP 33120A) that deliver a tone burst modified by a Hanning window function to the transducers. The central frequency is taken equal to 200 kHz. The incident Lamb wave of a specific mode is launched by means of two identical piezo-ceramic transducers (PZT-27) placed at the opposite sides of the plate edge. The thickness, the width and the length of these transducers are equal to 1, 6 and 50 mm, respectively. The free lateral resonance frequency of these transducers is near to 200 kHz. The generation of the A0 mode is obtained by exciting the two thin piezo-ceramic transducers with anti-phased electrical signals. On the contrary, the in-phased excitation generates the S0 mode. Two conventional Panametrics transducers are placed on the plate surface before and after the notch. Local honey and gel coupling are employed for the emitters and receivers, respectively. All the signals from the sensors are then recorded using a digital oscilloscope and transferred via the GPIB bus to a computer for signal processing. The aluminum plate used in the experimental investigations is considered with thickness (2d) and length (2L) equal to 6 and 500 mm, respectively. The longitudinal velocity (CL), the transverse velocity (CT) and the density (ρ) of this plate are equal to 6422, 3110 m/s and 2695 kg/m3, respectively. The plate is considered lossless. Here, the notch has a variable width equal to w. The plate thickness changes abruptly either from 2d to 2dp or from 2dp to 2d, with d the half thickness of the plate and p takes values from 0 to 1 with a constant increment.

In this work, the fundamental Lamb mode, S0 or A0, is launched from the plate edge. The generation of the S0 mode is performed by the application, at the left edge of the plate, of both tangential symmetrical and normal anti-symmetrical displacements, with respect to the median plane, windowed by Hanning temporal function. Alternatively, the generation of the A0 mode is obtained by the application of the tangential anti-symmetrical and the normal symmetrical displacements.

For the symmetrical notches, no mode conversion is observed, which is consistent with the theoretical symmetry principles. The figure 5 shows the comparison between the reflection and transmission power coefficients obtained when the A0 and S0 modes are launched. We can see a good agreement between the experimental and the numerical results.

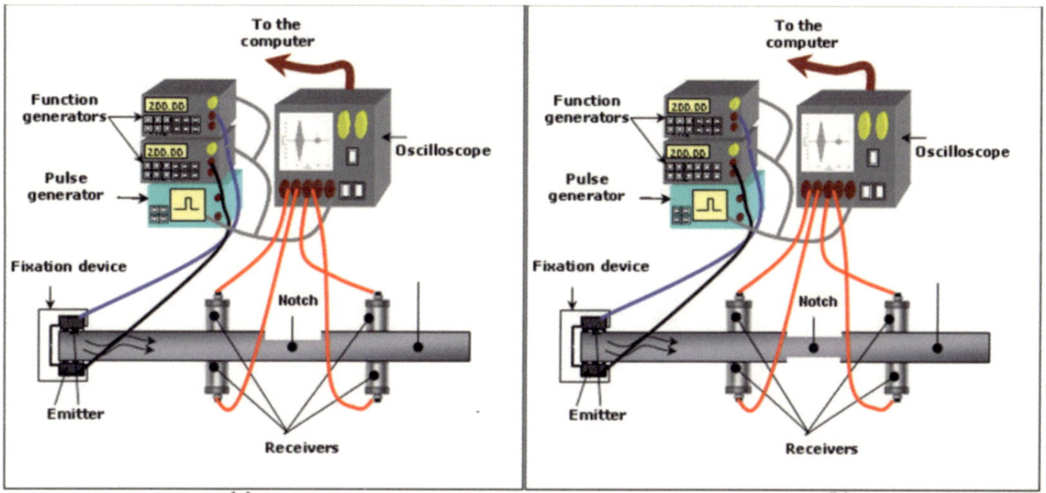

(a) (b)

Fig. 4. Experimental device to study the interaction of Lamb modes with (a) asymmetrical discontinuities and (b) symmetrical discontinuities.

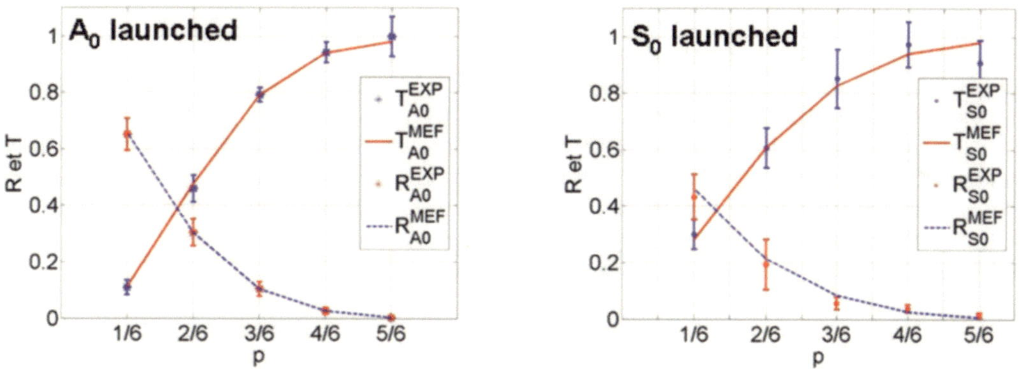

Fig. 5. Comparison between numerical and experimental results for the reflection (R) and the transmission (T) coefficients, when the launched Lamb mode A0 (a) or S0 (b) interacts with a symmetrical discontinuity.

For the asymmetrical notches a mode conversion from the incident mode A0 to the converted mode S0 and inversely is enabled. The curves of figure 6 show a good agreement between the experimental and the numerical results for the reflection coefficient, for both incident and converted modes, when the A0 or S0 mode is launched.

3.3 Dual signal processing for damage detection

The identification of Lamb mode amplitude variation as a function of the damage evolution is still the most difficult step in the process of damage monitoring using embedded Lamb wave-based systems. The aim of this section is to propose a simple system based on the generation of two different frequencies in order to better identify Lamb mode amplitude and to avoid false data interpretation in plates containing a hole of variable diameter. This

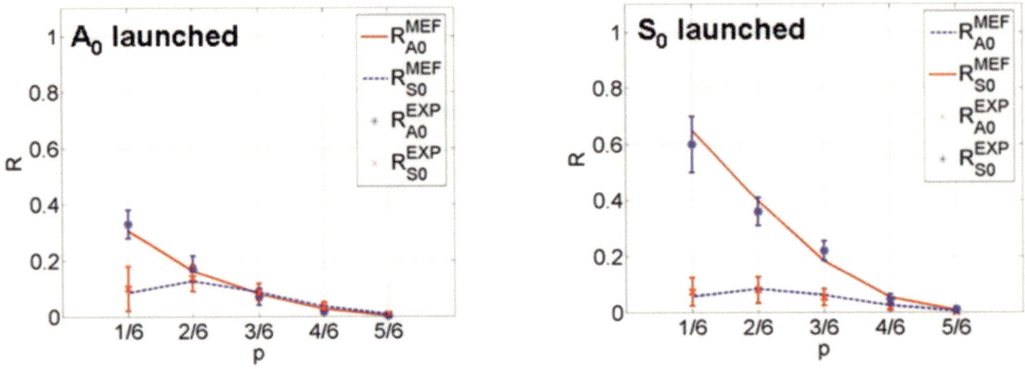

Fig. 6. Comparison between numerical and experimental results for the reflection (R) coefficient, when the launched Lamb mode A0 (a) or S0 (b) interacts with an asymmetrical discontinuity.

identification is based on a simple relation between the short-time Fourier transform (STFT) and the two-dimensional Fourier transform (2DFT).

Experimentally, a 3 mm thick aluminum plate with an emitter and a receiver is considered (Figure 7). The emitter consists of two piezoelectric elements with different widths in order to allow the excitation of two different frequency bands. Their sizes were equal to 15 × 3 × 1 mm^3 and 15 × 2 × 1 mm^3, which correspond to transverse resonance frequencies of 400 kHz and 600 kHz, respectively. The receiver consists of a 32-electrodes piezoelectric transducer with an inter-electrode distance of 2 mm, and 400 μm thick, 15 mm wide and 63 mm long. It was developed especially to increase the received power. The initial distance between the emitter and the sensor was chosen to be 25 cm. The Lamb waves generated by the emitter were received on different electrodes of the sensor. All the signals were then recorded by a digital oscilloscope with a sampling frequency equal to 5 MHz and transferred to a PC where the signal processing, i.e. STFT and 2DFT, could be applied.

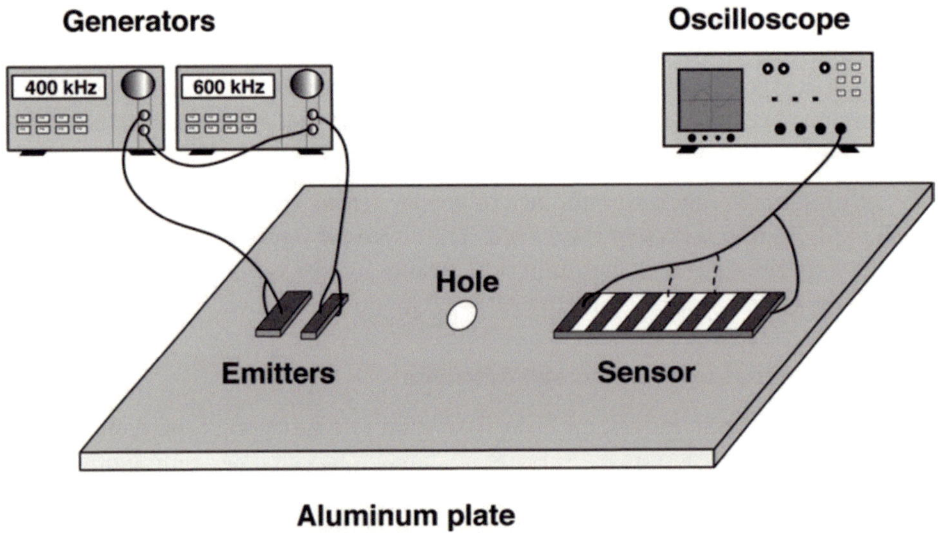

Fig. 7. Experimental measurement setup

In order to simulate the damage, a hole of variable diameter is introduced at the plate centre, and its diameter was increased from 1 mm to 13 mm.

By computing the amplitudes on the 2DFT and STFT analysis as functions of the diameter of the hole, the sensitivity of Lamb modes can be analyzed. In fact, the application of the dual signal processing approach to the received Lamb wave signals allows us to monitor the damage evolution using the *A*0 and *S*0 Lamb mode amplitude variation in two different frequency bands. Moreover, it is shown (figure 8) that for one of the frequencies a false interpretation can be induced when using only one signal processing technique.

It can be noted that both *S*0 and *A*0 modes are sensitive to the presence of the hole for both frequencies 400 kHz and 600 kHz, but the interaction of the same mode at different frequencies does not give similar results. At 600 kHz and for both *S*0 and *A*0 modes (see figures 8(*a*) and (*b*)), the amplitudes of the 2DFT and the STFT drop quasi-continuously according to the hole diameter by keeping close values. Both methods give results in good agreement and demonstrate the validity of their use. In contrast to this, in the case of the

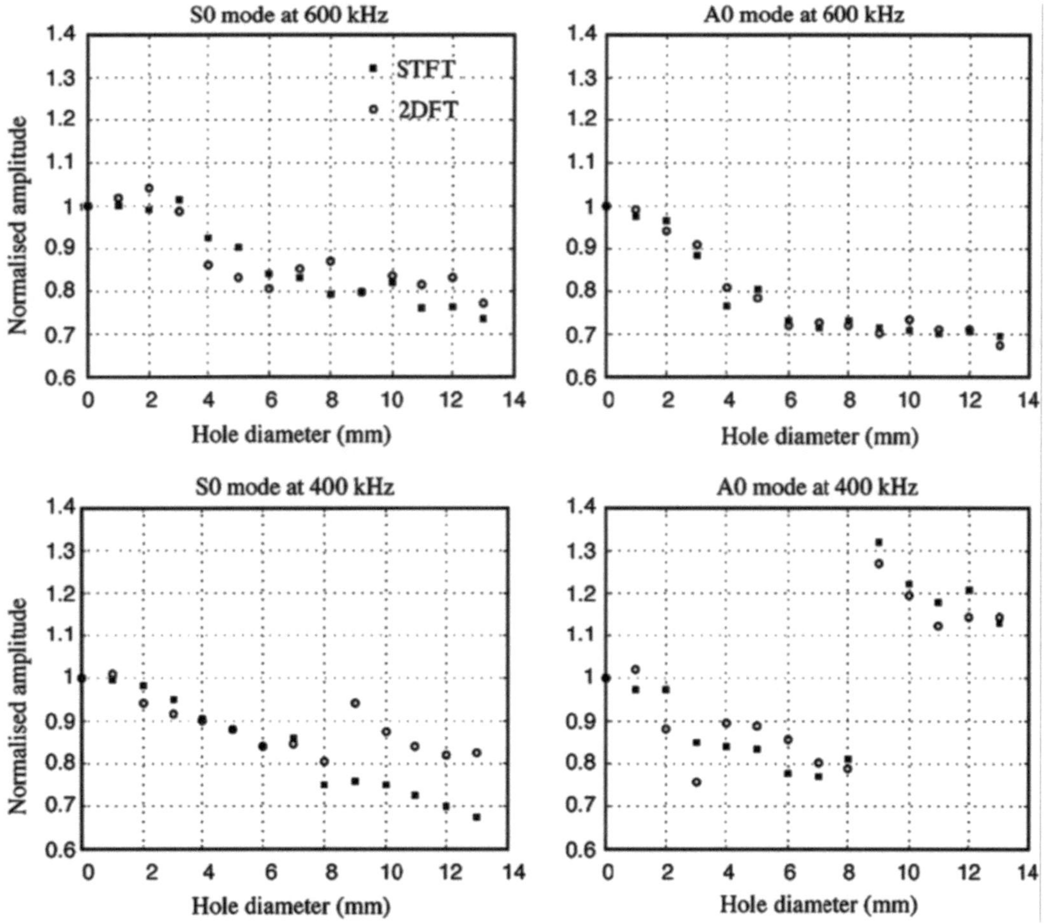

Fig. 8. STFT and 2DFT amplitude variation as a function of the hole diameter. (*a*) *S*0 mode at 600 kHz; (*b*) *A*0 mode at 600 kHz, (*c*) *S*0 mode at 400 kHz; (*d*) *A*0 mode at 400 kHz.

*S*0 mode at 400 kHz (see figure 8(*c*)) and for a diameter equal to approximately 9 mm, the amplitude of the 2DFT increases whereas the amplitude of the STFT decreases. Although this variation is relatively small, it represents typically a problem of false data interpretation. Simultaneously, in the case of the *A*0 mode at 400 kHz (see figure 8(*d*)), the STFT and the 2DFT amplitudes plotted as functions of the hole diameter show that the amplitudes are considerably increased for a hole diameter greater than 9 mm. Again a false data interpretation can be induced concerning the severity of the damage. This change can be related to a resonance phenomenon related to the presence of the hole. A better understanding of this phenomenon requires three-dimensional studies.

These measurements demonstrate the ability to use the STFT and 2DFT at the same time in order to detect damage and to overcome the problem of false data interpretation. In fact, using only one technique and one frequency can not always allow us to get the severity of the damage and consequently to determine the Lamb wave sensitivities to the presence of damage.

Although the results obtained with the STFT are more satisfactory than the 2DFT in this case, they would be more severe in error if the group velocities of the two modes were more similar or if the tested structure was more complex.

3.4 Passive SHM using ambient acoustic field cross-correlation

Recent theoretical and experimental studies have shown the possibility to retrieve the Green function between two points in a structure by cross-correlating the received signals at these points simultaneously, in the presence of a diffuse acoustic field in the medium. The aim of the work presented in this section is to exploit the mechanical vibrations and subsequent elastic wave fields present in an aeronautic structure during the flight. These vibrations being the result of the turbo engines and the aero acoustic phenomena, their random character makes the exploitation complicated. Meanwhile, the non need of an active source in this case, is a very interesting solution from an energy consumption point of view.

In the following, the reproducibility of the cross-correlation function, its potential to detect a defect, and its sensitivity to the source characteristics are studied. In fact, since the measured cross-correlation is used to monitor the integrity of the structure, it has to be reproducible for different measurements done in the same conditions. A second necessary condition to the study is the ability to detect any form of heterogeneity in the structure, using the cross-correlation of the ambient acoustic field. Finally, since the source-position influences the result, it is crucial to avoid misinterpretations by separating changes caused by source motion from those caused by defect appearance.

3.4.1 Reproducibility of the cross-correlation function

To study the applicability of the ambient noise correlation method, experimentation in the laboratory has been set-up in order to verify the reproducibility and the sensitivity of the correlation function to a defect. Thus, an aluminum plate of 2*1 m^2-surface and 6mm-thickness has been considered, and two circle PZ27-piezoelectric transducers of 0.5cm-radius and 1mm-thickness have been glued with honey at two positions *A* and *B*. To generate the ambient acoustic noise in the plate, an electrical noise generator has been used, and the signal has been emitted using a circle PZ27-piezoelectric transducer of 1cm-radius and 1mm-thickness, placed at a position *O*. The signals received at *A* and *B* have been measured and sent to a computer using a GPIB bus (Figure 9).

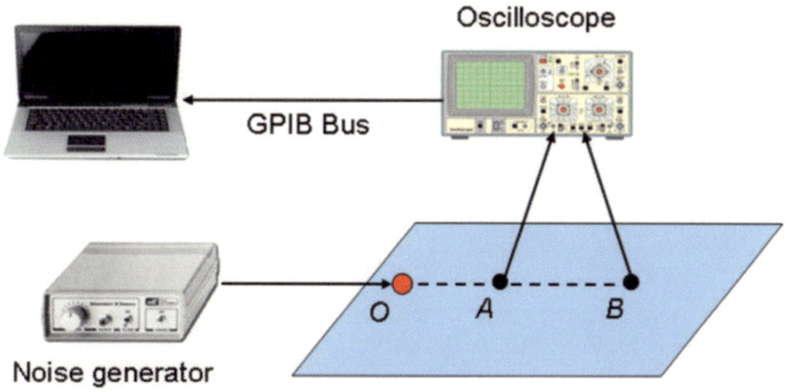

Fig. 9. Experimental set-up for studying reproducibility.

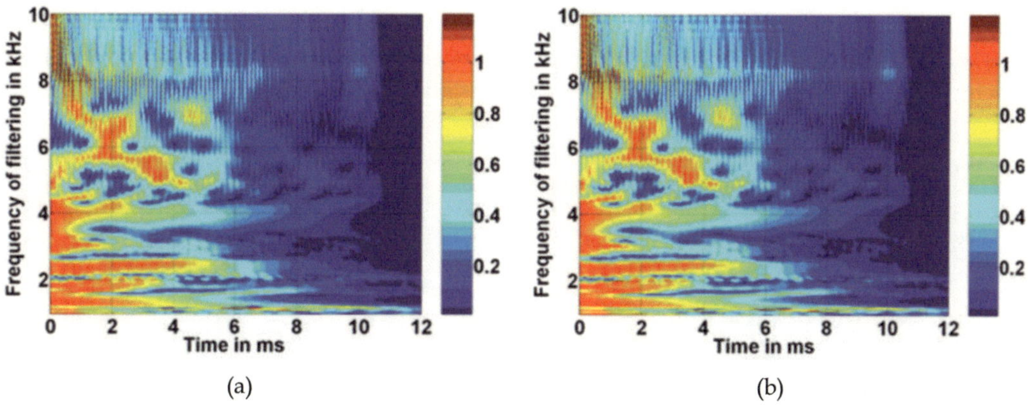

Fig. 10. Cross-correlation function between A and B for the measurement (a) #1 (b) #2.

The cross-correlation between these two signals has then been computed and averaged on $N = 150$ acquisitions to increase the signal-to-noise ratio. Finally, in order to better analyze the signals, a time-frequency representation has been used. Thus, a wavelet-transform of the measured cross-correlation function has been computed by convolving with a 5-cycle Hanning-windowed sinusoid of variable central frequency f_0. A time-frequency representation of the cross-correlation function, in the frequency-band [1-6 kHz], is shown at figure 10. We can see that for two measurements done in the same conditions, the cross-correlation function is reproducible.

3.4.2 Sensitivity of the cross-correlation function to a defect

In this section, the sensitivity of the cross-correlation function to a defect is studied. Thus, two measurements have been done, one without a defect and the other with a defect somewhere in the plate (Figure 11). Concerning the modeling of the defect, for repeatability purpose, an aluminum disk of 1 cm-radius has been glued on the surface of the plate between the two points *A* and *B*. In fact, such a defect introduces local heterogeneity from an acoustic impedance change point of view.

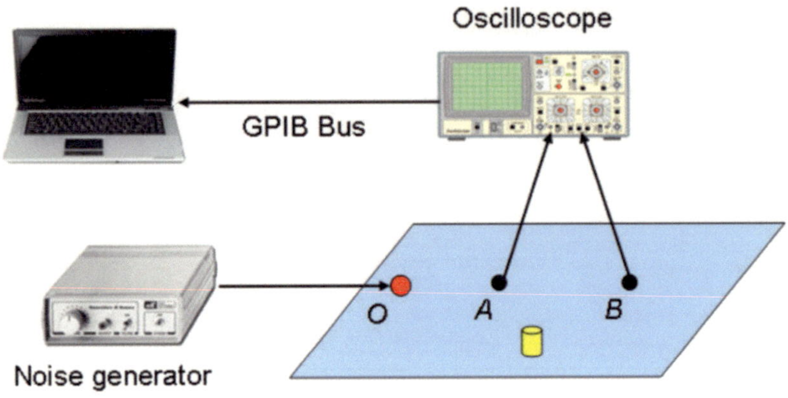

Fig. 11. Experimental set-up for defect detection.

The comparison of the measurements (Figure 12) with and without a defect shows that the cross-correlation function is sensitive to the presence of the defect. The sensitivity is more or less important depending on the frequency range and the position of the defect.

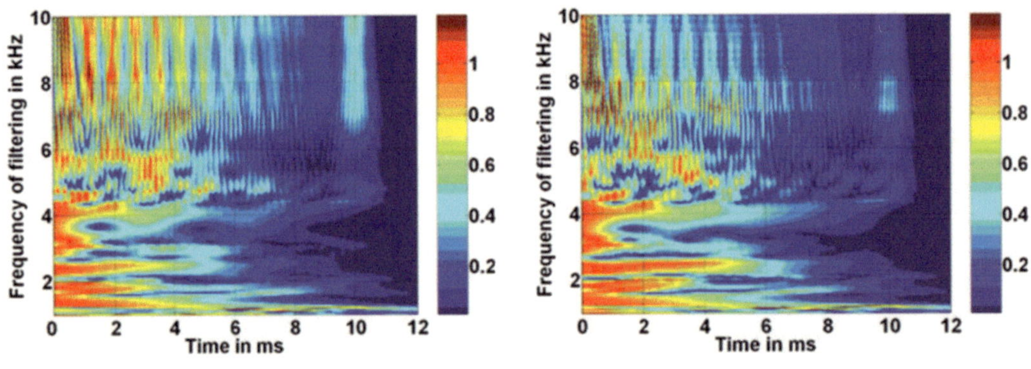

Fig. 12. Cross-correlation function between A and B (a) without and (b) with a defect.

3.4.3 Influence of the source characteristics on the cross-correlation function

The reproducibility and the sensitivity to a defect of the cross-correlation function being verified, this section will deal with the study of the influence of the source position on the correlation function. To better highlight the influence of the source position on the cross-correlation function, experimentation with two source positions is done. The images of the figure 13, for the first source position (figure 13.a) and the second source position (Figure 13.b), show that the cross-correlation function depends strongly on the source position. This influence of the source position could be misinterpreted as the appearance of a defect. A solution to this problem is given in the next section.

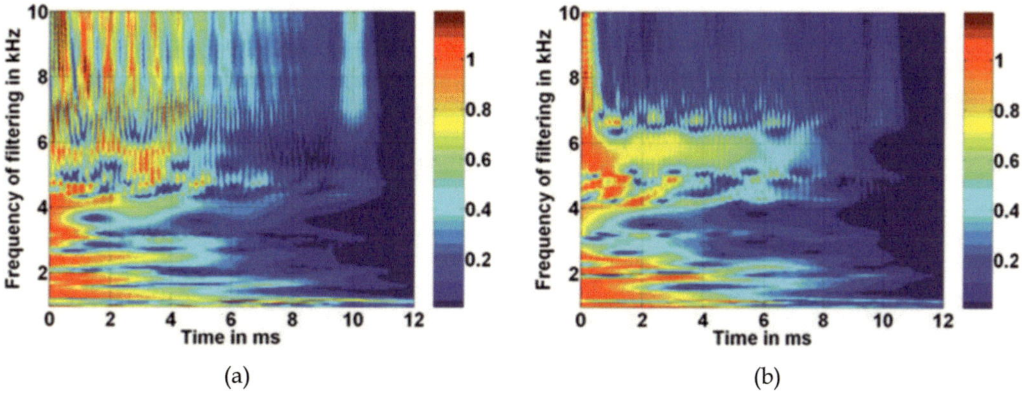

Fig. 13. Cross-correlation function between A and B (a) for the first and (b) for the second source position.

3.4.4 Practical application of the ambient noise correlation technique to SHM

In an aeronautic application, the sources exploited can be concentrated and with variable characteristics. This represents a major difficulty for the application. Indeed, in this situation, it is difficult to separate the contributions of the characteristics of the medium, from the characteristics of the source, in the measured information. To overcome this problem, we proposed a solution based on using a third transducer, called "reference transducer" and placed far from A and B, to identify the acoustic source characteristics at the instant of measurement by computing the auto-correlation of the received signal, before doing the diagnostic of the structure.

A simple experimentation has been set-up in the laboratory in order to test the applicability of the principle. Three piezoelectric receivers have been glued, at respective locations A, B and C, on an aluminum plate of 2 m × 1 m surface and 6 mm thickness (Fig. 14). The "ambient" acoustic noise is generated using an amplified loudspeaker working in the audible range (up to approximately 8 kHz), placed under the plate and driven by an electrical noise generator. High-pass filtering is applied in order to reject frequencies below 2 kHz. For repeatability purpose, a "removable" defect was used here instead of an actual structural damage: a small aluminum disk of 1 cm radius bonded between A and B.

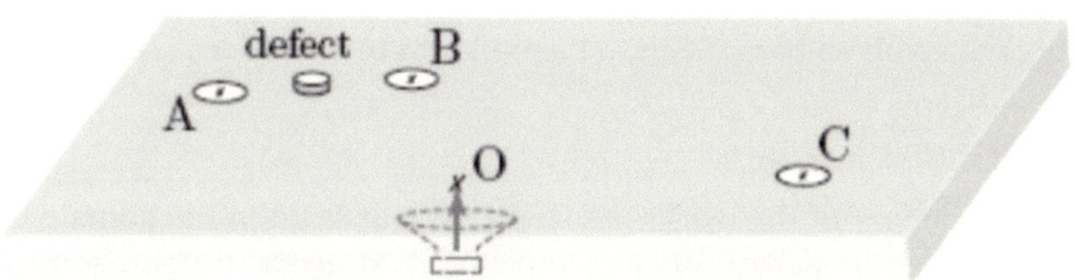

Fig. 14. Description of the experimental setup

The cross-correlation of 0.5 s-long signals measured at positions A and B, and the auto-correlation at position C have been averaged over 150 acquisitions. In order to emphasize interesting effects, narrowband filtering has been applied by convolving it with an N-cycle Hanning-windowed sinusoid of variable central frequency f_0.

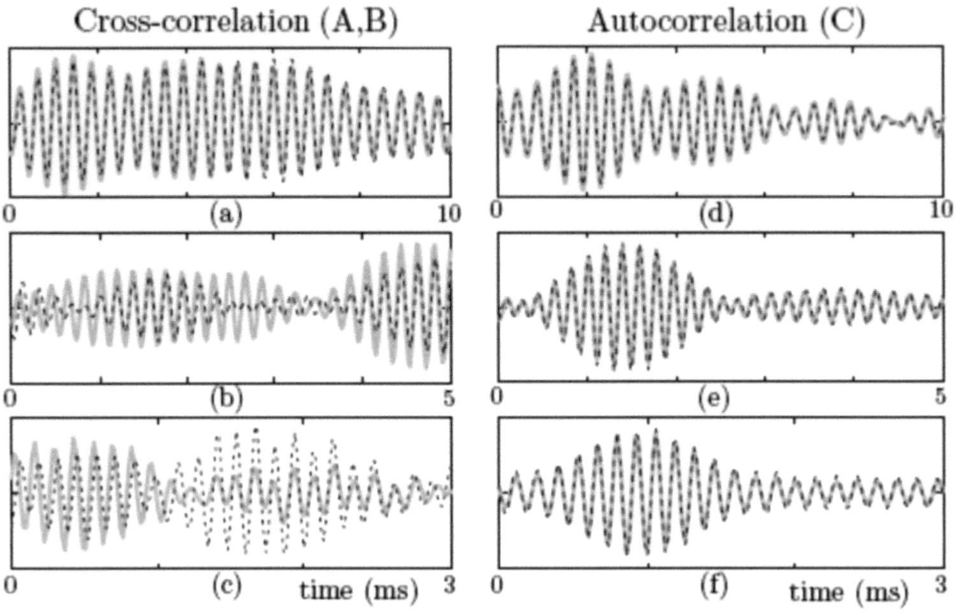

Fig. 15. Filtered average cross-correlation (a, b, c) and autocorrelation (d, e, f) functions with (broken line) and without (solid line) defect. (a), (d) f_0 = 2.5 kHz, N = 10 cycles. (b), (e) f_0 = 5.2 kHz, N = 15 cycles. (c), (f) f_0 = 7.8 kHz, N = 15 cycles.

Thus, comparisons of the results obtained in the absence and in the presence of defect are shown in Fig. 15 for three representative values of f_0. The curves (a), (b) and (c) show that except in the lower frequency case, the presence of the defect induces significant modifications of the cross-correlation function. As in a typical pitch-catch measurement, amplitude variations as well as phase shifts are observed. Contrariwise, the curves (d), (e) and (f) show that in the same conditions, the auto-correlation at the receiver C is unaffected by the presence of the defect near A and B.

The reliability of the proposed solution depends clearly of the position of the reference receiver C, which should be at the same time not sensitive to the appearance of a defect (in other words far from the inspection area), and sensitive to the source characteristics (close to the source). To quantify more precisely the sensibility of the autocorrelation to the defect and to identify the involved parameters, a theoretic study was done [37].

4. Conclusion

In this paper, a summary of the works developed by our team in the domain of SHM were presented. Thus, the modeling of a complete SHM system (emission, propagation, reception) using finite element method was shown. Then, the study on the interaction of Lamb waves with different types of discontinuities by calculating the power transmission and reflection coefficients was done. In order to better understand this interaction, a dual signal processing based on STFT and 2DFT was presented. This technique allows separating the influence of damage on each Lamb's mode. Finally, a new SHM technique based on the exploitation of the natural acoustic vibration in an aircraft during flight was shown which is very interesting from an energy consumption point of view. The feasibility of this method was experimentally validated by proposing a solution that allows separating the characteristics of the source and those of the medium. Encouraging results make possible considering the development of autonomous, integrated wireless network sensors for passive SHM application.

5. References

[1] D. C. Worlton, "Ultrasonics testing with lamb waves",Nondestruct. Test. 15, 218–222 (1957).

[2] D. N. Alleyne and P. Cawley, "The interaction of lamb waves with defects", IEEE Trans. Son. Ultrason. 39, 381–397 (1992).

[3] R. S. C. Monkhouse, P. D. Wilcox, and P. Cawley, "Flexible interdigital pvdf lamb wave transducers for the development of smart structures", Rev. Prog. Quant. Nondestruct. Eval. 16A, 877–884 (1997).

[4] J.-B. Ihn and F.-K. Chang, "Detection and monitoring of hidden fatigue crack growth using a built-in piezoelectric sensor/actuator network: I.diagnostics", Smart Mater. Struct. 13, 609–620 (2004).

[5] P. Blanquet, "Etude de l'endommagement des matériaux composites aéronautiques à partir de techniques ultrasonores", PhD Thesis, Report 9724, University of Valenciennes, France (1997).

[6] T .Demol, "Etude de transducteurs en barettes pour le contrôle santé des structures aéronautiques composites par ondes de Lamb. Application à la caractérisation de l'impact basse vitesse", PhD Thesis, Report 9801, University of Valenciennes, France (1998).

[7] E. Moulin, J. Assaad, C. Delebarre, H. Kaczmarek and D. Balageas, ''Piezoelectric transducer embedded in composite plate : Application to Lamb wave generation,'' J. Appl. Phys. vol. 82, pp 2049-2055 (1997).

[8] E. Moulin, J. Assaad, C. Delebarre et D. Osmont, ''Modelling of Lamb waves generated by integrated transducers in composite plates, using a coupled Finite Element - Normal Modes Expansion method,'' J. Acoust. Soc. Am, vol. 107, pp 87-94 (2000).

[9] E. Moulin, J. Assaad, C. Delebarre et S. Grondel, ''Modeling of integrated Lamb waves generation systems using a coupled finite element - normal modes expansion technique,'' Ultrasonics, vol. 38, pp 522-526 (2000).

[10] E. Moulin, S. Grondel, M. Baouahi, J. Assaad, "Pseudo-3D modeling of a surface-bonded Lamb wave source", J. Acoust. Soc. Am., vol. 119, pp 2575-2578, 2006.

[11] E. Moulin, S. Grondel, J. Assaad, L. Duquenne, "Modeling a surface-mounted Lamb wave emission-reception system : Applications to structural health monitoring", à paraître dans J. Acoust. Soc. Am., 2008.

[12] V. Giurgiutiu, "Tuned lamb wave excitation and detection with piezoelectric wafer active sensors for structural health monitoring", J. Intell. Mater. Syst. Struct. 16, 291–305 (2005).

[13] J. H. Nieuwenhuis, J. J. Neumann, D. W. Greve, and I. J. Oppenheim, "Generation and detection of guided waves using pzt wafer transducers", IEEE Trans. Ultrason. Ferroelectr. Freq. Control 52, 2103–2111 (2005).

[14] F. L. di Scalea, H. Matt, and I. Bartoli, "The response of rectangular piezoelectric sensors to raleigh and lamb ultrasonic waves", J. Acoust. Soc. Am. 121, 175–187 (2007).

[15] Grondel S, Delebarre C, Assaad J, Dupuis J P and Reithler L, " Fatigue crack monitoring of riveted aluminium strap joints by Lamb wave analysis and acoustic emission measurement techniques" NDT&E Int. 25 137–46 (2002)

[16] C Paget, S Grondel, K Levin, C Delebarre, "Damage assessment in composites by Lamb waves and wavelet coefficients - Smart Materials and Structures" 12, p. 393-402, 2003

[17] L. Duquenne, E. Moulin, J. Assaad, S. Grondel, "Transient modeling of Lamb waves generated in viscoelastic materials by surface bonded piezoelectric transducers", J. Acoust. Soc. Am, vol. 116, pp 133-141, 2004.

[18] El Youbi F, Grondel S and Assaad J. "Signal processing for damage detection using two different array transducers" Ultrasonics 42 803–6. (2004)

[19] F. Benmeddour, S. Grondel, J. Assaad, E. Moulin, "Study of the fundamental Lamb modes interaction with symmetrical notches", NDT\&E Int., vol. 41, pp 1-9, 2007.

[20] F. Benmeddour, S. Grondel, J. Assaad, E. Moulin, "Study of the fundamental Lamb modes interaction with asymmetrical discontinuities", NDT\&E Int., vol. 41, pp 330-340, 2008.

[21] M. Baouahi, "Modélisation 3D de la generation des ondes de Lamb par des transducteurs piézoéléctriques mono et multi-éléments", PhD Thesis, Report 0711, University of Valenciennes, France (2007).

[22] Diligent O, Grahn T, Bostro¨m A, Cawley P, Lowe MJS. "The low-frequency reflection and scattering of the S0 Lamb mode from a circular through-thickness hole in a plate: finite element, analytical and experimental studies." J Acoust Soc Am 2002;112(6):2589–601.

[23] Grahn T. "Lamb wave scattering from a circular partly through thickness hole in a plate." Wave Motion 2003;37:63–80.

[24] Guo N, Cawley P. "The interaction of Lamb waves with delaminations in composite laminates." J Acoust Soc Am 1993; 94(4):2240–6.

[25] Hayashi T, Kawashima K. "Single mode extraction from multiple modes of Lamb wave and its application to defect detection". JSME Int J 2003; 46(4):620–6.

[26] Le-Cle´zio E, Castaings M, Hosten B. "The interaction of the S0 Lamb mode with vertical cracks in an aluminum plate." Ultrasonics 2002; 40:187–92.

[27] Wang L, Shen J. "Scattering of elastic waves by a crack in a isotropic plate." Ultrasonics 1997; 35:451–7.

[28] Cho Y, Rose JL."An elastodynamic hybrid boundary element study for elastic wave interactions with a surface breaking defect." Int J Sol Struct 2000; 37:4103–24.

[29] Lowe MJS, Challis RE, Chan CW. "The transmission of Lamb waves across adhesively bonded lap joints." J Acoust Soc Amer 2000; 107(3):1333–45.

[30] Mal AK, Chang Z, Guo D, Gorman M. Lap-joint inspection using plate waves. In: Rempt RD, Broz AL (Eds.), SPIE nondestructive evaluation of aging aircraft, airports, and aerospace hardware, vol. 2945, 1996, p. 128–137.

[31] Cho Y. "Estimation of ultrasonic guided wave mode conversion in a plate with thickness variation." IEEE Trans Ultrason Ferroelectr Freq Control 2000;17(3):591–603.

[32] Alleyne DN, Cawley P." A 2-dimensional fourier transform method for the quantitative measurement of Lamb modes." IEEE Ultrason Symp 1990;1143–6.

[33] Alleyne DN, Cawley P. "The measurement and prediction of Lamb wave interaction with defects". IEEE Ultrason Sympos 1991; 855–7.

[34] Lowe MJS, Cawley P, Kao J-Y, Diligent O. "Prediction and measurement of the reflection of the fundamental anti-symmetric Lamb wave from cracks and notches". In: Thompson DO, Chimenti DE, editors. Review of progress in quantitative NDE, vol. 19A. New york: Plenum; 2000. p. 193–200.

[35] Lowe MJS, Cawley P, Kao J-Y, Diligent O. "The low frequency reflection characteristics of the fundamental antisymmetric Lamb wave A0 from a rectangular notch in a plate." J. Acoust Soc Am 2002; 112(6):2612–22.

[36] Lowe MJS, Diligent O." Low-frequency reflection characteristics of the S0 Lamb wave from a rectangular notch in a plate". J Acoust Soc Am 2002;111(1):64–74.

[37] E. Moulin, N. Abou Leyla, J. Assaad, and S. Grondel, "Applicability of acoustic noise correlation for structural health monitoring in nondiffuse field conditions", Appl. Phys. Lett. 95, 094104 (2009).

[38] N. M. Shapiro, M. Campillo, L. Stehly, and M. Ritzwoller. "High-resolution surface-wave tomography from ambient seismic noise", Science 29, 1615-1617 (2005)

[39] K. Wapenaar, "Retrieving the elastodynamic Greens function of an arbitrary inhomogeneous medium by cross correlation", Phys. Rev. Lett. 93, 254301 (2004)

[40] K. G. Sabra, P. Gerstoft, P. Roux, W. Kuperman, and M. C .Fehler "Surface wave tomography using microseisms in southern california", Geophys. Res. Lett. 32, L023155 (2005)

[41] P. Roux, W. A. Kuperman, and the NPAL Group, "Extracting coherent wavefronts from acoustic ambient noise in the ocean", J. Acoust. Soc. Am. 116, 1995-2003 (2004)

[42] K. G. Sabra, P. Roux, W. A. Kuperman "Arrival-time structure of the time-averaged ambient noise cross-correlation function in an oceanic waveguide", J. Acoust. Soc. Am. 117, 164-174 (2005)

[43] K. G. Sabra, E. S. Winkel, D. A. Bourgoyne, B. R. Elbing, S. L. Ceccio, M. Perlin, and D. R. Dowling "Using cross-correlations of turbulent flow-induced ambient vibrations to estimate the structural impulse response. Application to structural health monitoring", J. Acoust. Soc. Am. 121, (4) (2007)

[44] K. G. Sabra A. Srivastava, F. Lanza Di Scalea, I. Bartoli, P. Rizzo and S. Conti, "Structural health monitoring by extraction of coherent guided waves from diffuse fields", J. Acoust. Soc. Am. 123, (1) (2008)

[45] E. Larose, P. Roux, and M. Campillo, "Reconstruction of Rayleigh-Lamb dispersion spectrum based on noise obtained from an air-jet forcing", J. Acoust. Soc. Am. 122, (6) (2007)

[46] L. Duquenne, F. Elyoubi, E. Moulin, S. Grondel, J. Assaad, and C. Delebarre, "The use of permanently-mounted surface transducers to characterize lamb wave propagation", in *Proc. World Congress Ultras.*, 601-604 (Paris, France) (2003).

6

Piezoelectric Transducers Applied in Structural Health Monitoring: Data Acquisition and Virtual Instrumentation for Electromechanical Impedance Technique

Fabricio Guimarães Baptista and Jozue Vieira Filho
Department of Electrical Engineering, Sao Paulo State University, Ilha Solteira, Brazil

1. Introduction

This chapter reports the application of piezoelectric transducers in the detection of structural damage, focusing on the data acquisition based on virtual instrumentation for the appropriate analysis of the signals from these transducers, allowing fast and accurate measurements.

There is a growing interest in systems able to continuously monitor a structure and timely detect incipient damage, ensuring a high level of safety and reducing maintenance costs. This concept is commonly known as Structural Health Monitoring (SHM) in the literature. Many techniques can be used to develop SHM systems, but the electromechanical impedance (EMI) technique has the advantage of using small and very thin piezoelectric patches, with thickness on the order of a few tenths of millimeters. Thus, these devices are like adhesives bonded to the monitored structure, allowing a large area of the structure to be monitored with negligible effects on its mechanical properties. Due to the piezoelectric effect, when a piezoelectric transducer is bonded to the structure to me monitored, there is an interaction between the mechanical impedance of the host structure and the electrical impedance of the transducer. Therefore, changes in the mechanical impedance of the host structure caused by damage, such as cracks or corrosions, can be detected simply by measuring the electrical impedance of the transducer in a suitable frequency range, which is easier to measure than the mechanical impedance.

Usually, the measurement of the electrical impedance of piezoelectric transducers, basic stage of the technique, is carried out by commercial impedance analyzers. Although accurate, these instruments are expensive, bulky and slow for real-time SHM applications, where fast measurements from multiple sensors are required. According to the results presented here, the virtual instrumentation can be a great ally in the development of real-time SHM systems in structures with large number of piezoelectric transducers. Measurement systems based on virtual instrumentation are fast and extremely versatile, allowing adjustments to be easily incorporated in accordance with the user needs.

This chapter is organized as follows. A brief introduction to SHM is presented in Section 2, indicating its importance, the main fields of application and the main methodologies involved. The EMI technique is discussed in Section 3. In this section, an equivalent electromechanical circuit is obtained to relate the electrical impedance of the transducer to the mechanical impedance of the monitored structure. Measurements systems based on virtual instrumentation are studied in Section 4. Two measurement methods are analyzed: frequency domain and time domain measurement. Finally, the chapter concludes with Section 5 citing the most relevant points.

2. Structural health monitoring (SHM)

This section is based on the literature review conducted by Sohn et al. (2004) from Los Alamos National Laboratory. The SHM systems are designed to continuously monitor a structure and to detect incipient structural damage. According to Rytter (1993), in advanced systems there is a five-step process to be followed in the characterization of damage: (1) damage detection; (2) location of damage in the structure; (3) determining which type of damage is present; (4) estimate its severity; (5) analysis of the remaining useful life of the structure, i.e., the prognosis.

Among the various fields of application, we can cite the large civil infrastructures (bridges, buildings, roads, oil rigs, etc.), the aeronautical and aerospace structures (aircraft, helicopters, satellites, space stations, etc.), and large marine structures (submarines and ships). There are both scientific and economic motivations for the use of an SHM system. From the scientific point of view, the monitoring and detection of structural damage mean to achieve a high level of safety. From the economic point of view, systems with this capability allow a significant reduction in maintenance costs. For example, Cawley (1997) suggested the use of an SHM system to identify corrosion in pipelines of chemical and petrochemical industries, in which the costs associated with the removal of these pipelines for inspection is prohibitive. The Federal Highway Administration estimates that nearly 35% of all bridges in the United States are either structurally or functionally deficient (Wang et al., 1997). The cost of repair or rebuilding lies in the billions of dollars Therefore, SHM systems could reduce this cost while providing high level of safety for users during repair or assessment.

Currently, the aviation industry is one of most focused fields of application. Although the design and criterions for certification of an aircraft already guarantee a high level of security, an SHM system could significantly reduce the repair and maintenance costs, which represent 27% of the cost of its life cycle (Kessler et al., 2002). The direct costs related to the repair could be reduced by detecting damage in an early stage. In addition, the indirect costs could be reduced by a lower frequency at which the aircraft would be shut down for maintenance.

The definition of damage is important in SHM systems. Damage is any change in the structure that may affect its performance and safety. Implicit in this definition is the concept that damage detection is based on the comparison between two states of the structure. In a first state, the structure is considered healthy and it is used as a reference for comparison with an updated state after a probable occurrence of damage. Damage such as cracks and corrosion can change various properties of the monitored structure as, for example, mass, stiffness, energy dissipation, mechanical impedance and its cross-section area. In addition, these changes alter the dynamic properties of the structure. From this concept, there are the techniques based on the frequency response function (FRF). In these techniques, the data for the evaluation of the structure are collected while the structure is in dynamic condition using one of two methods: ambient excitation and forced-excitation. In the ambient excitation, the vibration produced by the structure during its normal operation is utilized. The vibration of a bridge due to the vehicular traffic is an example of ambient excitation. On the other hand, the forced-excitation is controlled using actuators, such as shakers and piezoelectric transducers.

There are many techniques for developing SHM systems. Several techniques are based on non-destructive evaluation (NDE) methods, such as acoustic emission, comparative vacuum, magnetic particle inspection, Eddy current, and methods based on optical fibre sensors. In this chapter, we aim to analyze the electromechanical impedance (EMI) technique, which is a method based on forced-excitation. In this technique, low-cost and thin piezoelectric patches are bonded to the host structure and combine both the functions of actuator and sensor. The EMI technique is presented in the next section.

3. Electromechanical Impedance (EMI)

3.1 Basic concept

The EMI technique is a form of nondestructive evaluation based on the FRF which has the advantages of its simplicity and using thin and low-cost piezoelectric transducers. The most widely used piezoelectric transducers are the PZT (Pb-lead Zirconate Titanate) ceramics and MFC (Macro-Fiber Composite), which have thickness on the order of a few tenths of millimeters. These characteristics make the EMI technique especially attractive for the monitoring of aircraft structures, which is one of the most prominent application fields nowadays. The EMI technique is the only PZT-based technique that has characteristics for the development of a real-time and in-situ SHM system in aircrafts (Gyekenyesi et al., 2005). These devices are bonded to the structure to be monitored through a high-strength adhesive that can be instantaneous glue based on cyanoacrylate or epoxy resin. Due to the piezoelectric effect, there is a relationship between the mechanical properties of the structure and the electrical impedance of the transducer. Therefore, it is possible to monitor variations in these mechanical properties by measuring the electrical impedance. For example, Figure 1 shows a PZT patch from Piezo Systems bonded to an aluminum beam.

Fig. 1. A PZT patch bonded to an aluminum beam for damage detection.

In Figure 1, both sides of the PZT ceramic are coated by a thin metal layer, where wires are soldered to connection. Through these wires, the patch is excited and, simultaneously, its electrical impedance is measured in an appropriate frequency range. As a result, these ceramics combine the functions of sensor (direct piezoelectric effect) and actuator (reverse piezoelectric effect.

In order to obtain the relationship between the electrical impedance of the transducer and the mechanical impedance of the structure, we should analyze the wave propagation in the structure when the transducer is excited. For this analysis we consider the representation of a square PZT patch bonded to a host structure, as shown in Figure 2.

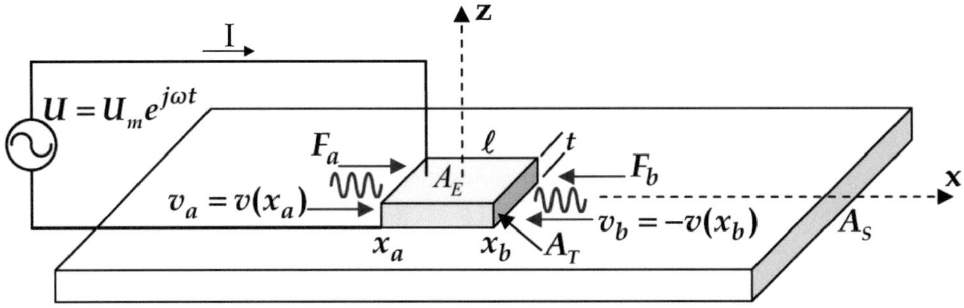

Fig. 2. Principle of the EMI technique; a square PZT patch is bonded to the structure to be monitored.

In Figure 2, a square PZT patch with side ℓ and thickness t is bonded to a rectangular structure with cross-sectional area A_S, which is perpendicular to the direction of its length. An alternating voltage U is applied to the transducer through the bottom and top electrodes, and the response is a current with intensity I. If the PZT patch has small thickness, a wave propagating at velocity v_a in the host structure reaches the patch side with coordinate x_a and surface area A_T causing the force F_a. Similarly, in the side of coordinate x_b there is a force F_b due to the incoming wave propagating at velocity v_b.

To find an equivalent circuit that represents the behavior of the PZT patch bonded to the structure, we need to determine the relationship between the mechanical quantities (F_a, F_b, v_a, v_b) and the electrical quantities (U, I), as shown in the next section.

3.2 Theoretical analysis

The theory developed in this section is based on analysis presented by Royer & Dieulesaint (2000).

As the thickness of the transducer is much smaller than the other dimensions, the deformation in its thickness direction (z-axis) due to the applied electric field is negligible. In general, for PZT patches of 5A and 5H type with thickness ranging from 0.1 to 0.3 mm, the deformation in the thickness direction is on the order of nanometers. On the other hand, the deformations in the sides ℓ (transverse direction) are on the order of micrometers. Therefore, the vibration mode is predominantly transverse to the direction of the applied electric field. In addition, if the applied voltage U is low in the order of a few volts and hence the resultant electric field is also low, the piezoelectric effect is predominantly linear and the non-linearities can be neglected. From this assumption and considering the class *6mm* for PZT ceramics (Meitzler, 1987), the basic piezoelectric equations for this case are given by

$$D_3 = d_{31}(T_1 + T_2) + d_{33}T_3 + \varepsilon_{33}^T E_3 \tag{1}$$

$$S_1 = s_{11}^E T_1 + s_{12}^E T_2 + s_{13}^E T_3 + d_{31}E_3 \tag{2}$$

$$S_2 = -s_{12}^E T_1 - s_{11}^E T_2 + s_{13}^E T_3 + d_{31}E_3 \tag{3}$$

where E_3 and D_3 are the electric field and electrical displacement, respectively; T_1, T_2, and T_3 are the *stress* components; S_1 and S_2 are the *strain* components; d_{31} and d_{33} are the piezoelectric constants; s_{11}, s_{12}, and s_{13} are the compliance components at constant electric field; ε_{33} is the permittivity at constant stress. The superscripts *E* and *T* donate constant electric field and constant stress, respectively, and the subscripts *1, 2,* and *3* refer to the directions *x, y,* and *z,* respectively.

Although the transducer is square and the deformations in both sides are approximately the same, only the deformation along the length of the structure is considered for the one-dimensional (1D) assumption. Thus, the main propagation direction is considered along the length direction (x-axis) perpendicular to the cross-section area A_S of the host structure, as shown in Figure 2. Therefore, for 1-D assumption, it is correct to consider $T_2 = T_3 = S_2 = 0$. Hence, the Equations (1) to (3) can be rewritten as follows

$$D_3 = d_{31}T_1 + \varepsilon_{33}E_3 \tag{4}$$

$$S_1 = s_{11}T_1 + d_{31}E_3 \tag{5}$$

The patch is essentially a capacitor. Thus, due to the voltage source, there is a charge density (ρ_e) on the electrodes of the patch and according to the Poisson equation we have

$$\frac{\partial D_3}{\partial z} = \rho_e \tag{6}$$

This results in a current of intensity $I_C e^{j\omega t}$. If the current is uniform over the entire area of the electrodes, the charge conservation requires

$$\frac{\partial D_3}{\partial t} = J(t) = \frac{I_C e^{j\omega t}}{A_E} \tag{7}$$

where $J(t)$ is the current density and A_E is the area of each electrode.

It is appropriate to put the stress T_1 in function of the electric displacement D_3. So, from Equation (5) and considering the following relation

$$S_x = \frac{\partial u_x}{\partial x} \tag{8}$$

where u_x is the displacement in the x direction, we can obtain

$$T_1 = \frac{1}{s_{11}}\frac{\partial u_x}{\partial x} - \frac{d_{31}}{s_{11}\varepsilon_{33}}D_3 \tag{9}$$

Differentiating Equation (9) with respect to time

$$\frac{\partial T_1}{\partial t} = \frac{1}{s_{11}}\frac{\partial}{\partial x}\left(\frac{\partial u_x}{\partial t}\right) - \frac{d_{31}}{s_{11}\varepsilon_{33}}\frac{\partial D_3}{\partial t} \tag{10}$$

and considering the velocity given by

$$v = \frac{\partial u_x}{\partial t} \tag{11}$$

and considering the charge conservation in Equation (7), we can rewrite the expression as follows

$$\frac{\partial T_1}{\partial t} = \frac{1}{s_{11}}\frac{\partial v}{\partial x} - \frac{d_{31}}{s_{11}\varepsilon_{33}}\frac{I_C e^{j\omega t}}{A_E} \tag{12}$$

The motion equation for this case is given by

$$\rho_T \frac{\partial^2 v}{\partial t^2} = \frac{1}{s_{11}}\frac{\partial^2 v}{\partial x^2} \tag{13}$$

where ρ_T is the mass density of the piezoelectric material. The general solution for Equation (13) is the sum of two waves propagating in opposite directions, as shown in Figure 2. In steady state, we have

$$v = (m \cdot e^{-jkx} + n \cdot e^{jkx})e^{j\omega t} = (v_m + v_n)e^{j\omega t} \tag{14}$$

where m and n are constants and k is the wave number given by

$$k = \frac{\omega}{V} \tag{15}$$

where V is the velocity of propagation given by

$$V = \frac{1}{\sqrt{s_{11}\rho_T}} \tag{16}$$

Substituting the velocity given in Equation (14) into the stress in Equation (12) and integrating with respect to time, we have

$$T_1 = \frac{1}{s_{11}}\frac{\partial}{\partial x}(m \cdot e^{-jkx} + n \cdot e^{jkx})\int e^{j\omega t}\mathrm{d}t - \frac{d_{31}}{s_{11}\varepsilon_{33}}\frac{I_C}{A_E}\int e^{j\omega t}\mathrm{d}t \tag{17}$$

$$T_1 = -\frac{k}{\omega s_{11}}(m \cdot e^{-jkx} - n \cdot e^{jkx})e^{j\omega t} - \frac{d_{31}}{j\omega s_{11}\varepsilon_{33}}\frac{I_C}{A_E}e^{j\omega t} \tag{18}$$

The characteristic (acoustic) impedance (Z_T^A) of the PZT patch is given by

$$Z_T^A = \frac{k}{\omega s_{11}} \tag{19}$$

Substituting Equation (19) into Equation (18) and hiding the term $e^{j\omega t}$ just for simplicity, the equation of stress can be rewritten as

$$T_1 = -Z_T^A(m \cdot e^{-jkx} - n \cdot e^{jkx}) + \mathrm{j}\frac{d_{31}}{s_{11}\varepsilon_{33}}\frac{I_C}{\omega A_E} \tag{20}$$

The forces acting on each face of the transducer can be calculated by

$$F_a = -A_T T_1(x_a) \tag{21}$$

$$F_b = -A_T T_1(x_b) \tag{22}$$

Thus, replacing Equation (20) into Equations (21) and (22) and considering the mechanical impedance of the transducer given by

$$Z_T = A_T Z_T^A = \frac{k}{\omega s_{11}} A_T \tag{23}$$

the forces F_a and F_b can be obtained as follows

$$F_a = Z_T(m \cdot e^{-jkx_a} - n \cdot e^{jkx_a}) - \mathrm{j}\frac{d_{31}}{s_{11}\varepsilon_{33}}\frac{A_T}{\omega A_E} I_C \tag{24}$$

$$F_b = Z_T(m \cdot e^{-jkx_b} - n \cdot e^{jkx_b}) - \mathrm{j}\frac{d_{31}}{s_{11}\varepsilon_{33}}\frac{A_T}{\omega A_E} I_C \tag{25}$$

The velocities v_a and v_b that reach the sides of the transducer with coordinates x_a and x_b, respectively, are given by

$$v_a = v(x_a) = m \cdot e^{-jkx_a} + n \cdot e^{jkx_a} \tag{26}$$

$$v_b = -v(x_b) = -m \cdot e^{-jkx_b} - n \cdot e^{jkx_b} \tag{27}$$

Considering the trigonometric identify $2jsin(\theta) = e^{j\theta} - e^{-j\theta}$ and the relation $x_b - x_a = \ell$, as shown in Figure 2, the terms m and n in Equations (26) and (27) can be computed as follows

$$m = \frac{v_a e^{jkx_b} + v_b e^{jkx_a}}{2jsin(k\ell)} \tag{28}$$

$$n = -\frac{v_a e^{-jkx_b} + v_b e^{-jkx_a}}{2jsin(k\ell)} \tag{29}$$

Replacing Equations (28) and (29) into Equations (24) and (25) and considering the trigonometric identify $2\cos(\theta) = e^{j\theta} + e^{-j\theta}$, the expressions for the forces F_a and F_b can be rewritten as

$$F_a = Z_T\left(\frac{v_a}{j\tan(k\ell)} + \frac{v_b}{jsin(k\ell)}\right) - j\frac{d_{31}}{s_{11}\varepsilon_{33}}\frac{A_T}{\omega A_E}I_C \tag{30}$$

$$F_b = Z_T\left(\frac{v_a}{jsin(k\ell)} + \frac{v_b}{j\tan(k\ell)}\right) - j\frac{d_{31}}{s_{11}\varepsilon_{33}}\frac{A_T}{\omega A_E}I_C \tag{31}$$

We need to determine the total current, which is the response of the transducer due to changes in the mechanical properties of the monitored structure. The total current can be obtained from the electric displacement.

The electric displacement in Equation (4) can be rewritten as

$$D_3 = \frac{d_{31}}{s_{11}}\frac{\partial u_x}{\partial x} + \varepsilon_{33}E_3 \tag{32}$$

The electric charge Q can be obtained from the electric displacement integrating Equation (32) with respect to area of the electrodes

$$Q = \int_S D_3 ds = \varepsilon_{33}E_3A_E + \frac{d_{31}}{s_{11}}\ell\left[u(x_b) - u(x_a)\right] \tag{33}$$

Since the PZT patch is very thin, the electric field is practically constant in the z-axis direction and it can be calculated as follows

$$E_3 = \frac{U}{t} \tag{34}$$

In addition, the static capacitance C_0 of the patch is given by

$$C_0 = \varepsilon_{33}\frac{A_E}{t} \tag{35}$$

Substituting Equations (34) and (35) into Equation (33), we obtain

$$Q = C_0U + \frac{d_{31}}{s_{11}}\ell\left[u(x_b) - u(x_a)\right] \tag{36}$$

Therefore, the total current (I_T) is obtained differentiating Equation (36) with respect to time, as fallows

$$I_T = \frac{dQ}{dt} = j\omega C_0U + \frac{d_{31}}{s_{11}}\ell j\omega\left[u(x_b) - u(x_a)\right] \tag{37}$$

The velocities v_a and v_b can be written as a function of the displacements $u(x_a)$ and $u(x_b)$, according to following equations

$$v_a = \frac{\partial}{\partial t}u(x_a) = j\omega u(x_a) \tag{38}$$

$$v_b = -\frac{\partial}{\partial t}u(x_b) = -j\omega u(x_b) \tag{39}$$

From Equations (38) and (39) we obtain

$$I_T = j\omega C_0 U - \frac{d_{31}}{s_{11}}\ell(v_a + v_b) = I_C - \frac{d_{31}}{s_{11}}\ell(v_a + v_b) \tag{40}$$

According to Equation (40), besides the current I_C due to the capacitance C_0, there is a current related to the velocities v_a and v_b.

Thus, the voltage U at the terminals of the transducer is given by

$$U = \frac{d_{31}}{j\omega C_0 s_{11}}\ell(v_a + v_b) + \frac{I_T}{j\omega C_0} \tag{41}$$

Or we can also write

$$U = \frac{I_C}{j\omega C_0} \tag{42}$$

Finally, equations (30), (31) and (42) can be rewritten in matrix form, as follows

$$\begin{pmatrix} F_a \\ F_b \\ U \end{pmatrix} = -j \begin{pmatrix} \frac{Z_T}{\tan(k\ell)} & \frac{Z_T}{sin(k\ell)} & \frac{d_{31}}{s_{11}\varepsilon_{33}}\frac{A_T}{\omega A_E} \\ \frac{Z_T}{sin(k\ell)} & \frac{Z_T}{\tan(k\ell)} & \frac{d_{31}}{s_{11}\varepsilon_{33}}\frac{A_T}{\omega A_E} \\ 0 & 0 & \frac{1}{\omega C_0} \end{pmatrix} \begin{pmatrix} v_a \\ v_b \\ I_C \end{pmatrix} \tag{43}$$

The matrix in Equation (43) is known as the electromechanical impedance matrix and defines the piezoelectric transducer as a hexapole, as shown in Figure 3.

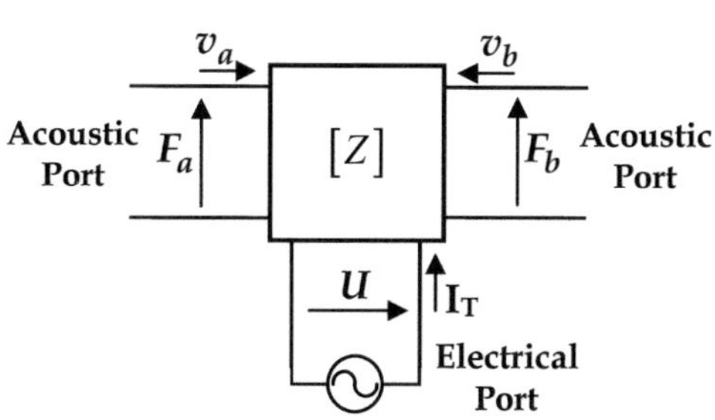

Fig. 3. Piezoelectric transducers can be represented as a hexapole with one electrical port and two acoustic ports.

According to Figure 3, the transducer is represented by a hexapole with one electrical port and two acoustic ports. Therefore, there is an electromechanical coupling with the monitored structure. Through the acoustic ports the structure is excited so that the dynamic properties can be assessed. Any variation in the dynamic properties of the structure caused by damage changes the mechanical quantities (F_a, F_b, v_a, v_b) and, due to the electromechanical coupling, also changes the electrical quantities (U, I). Therefore, the structural health can be monitored by measuring the current (I_T) and voltage (U) of the transducer.

In practice, the electrical impedance of the transducer is measured. The electrical impedance (Z_E) of the transducer is given by

$$Z_E = \frac{U}{I_T} \tag{44}$$

Thus, it is useful to find an equivalent electromechanical circuit that relates the electrical impedance of the transducer to the mechanical properties of the structure. The equivalent circuit is presented in the next section.

3.3 Equivalent electromechanical circuit

An electromechanical circuit makes it easy to analyze the electrical impedance of the transducer in relation to the dynamic properties of the structure, which are directly related to its mechanical impedance. Thus, we should obtain a circuit that establishes a relationship between the electrical impedance of the transducer and the mechanical impedance of the host structure.

Given the following trigonometric identify

$$\frac{1}{\tan(k\ell)} = \frac{1}{sin(k\ell)} - \tan\left(\frac{k\ell}{2}\right) \tag{45}$$

And considering the following manipulation

$$-j\frac{A_T}{\omega\varepsilon_{33}A_E}I_C = \ell\frac{t}{j\omega\varepsilon_{33}A_E}I_C = \ell\frac{1}{j\omega C_0}I_C = \ell U \tag{46}$$

We can rewrite Equations (30) and (31) as follows

$$F_a = -j\frac{Z_T}{sin(k\ell)}(v_a + v_b) + jZ_T \tan\left(\frac{k\ell}{2}\right)v_a + \frac{d_{31}}{s_{11}}\ell U \tag{47}$$

$$F_b = -j\frac{Z_T}{sin(k\ell)}(v_a + v_b) + jZ_T \tan\left(\frac{k\ell}{2}\right)v_b + \frac{d_{31}}{s_{11}}\ell U \tag{48}$$

From Equations (41), (42), (47), and (48), we can easily obtain the circuit shown in Figure 4 (a). The mechanical and electrical quantities are related through the electromechanical transformer with ratio (TR) given by

$$TR = \frac{d_{31}}{s_{11}}\ell \tag{49}$$

The circuit in Figure 4 (a) is not suitable for analysis of structural damage detection because it does not consider the monitored structure as a propagation media in each acoustic port of the transducer. Both the sides of the circuit, which corresponds to the acoustic ports, must be loaded by the mechanical impedance of the structure, as shown in Figure 4 (b).

The mechanical impedance (Z_S) is given by (Kossoff, 1966)

$$Z_S = A_S\left[\frac{\rho_S v}{1+\gamma^2} + j\frac{\gamma\rho_S v}{1+\gamma^2}\right] \tag{50}$$

where ρ_S is the mass density of the structure, A_S is the cross-sectional area of the structure, as shown in Figure 2, orthogonal to the wave propagating at velocity v and γ is the damping, i.e., the loss factor in nepers.

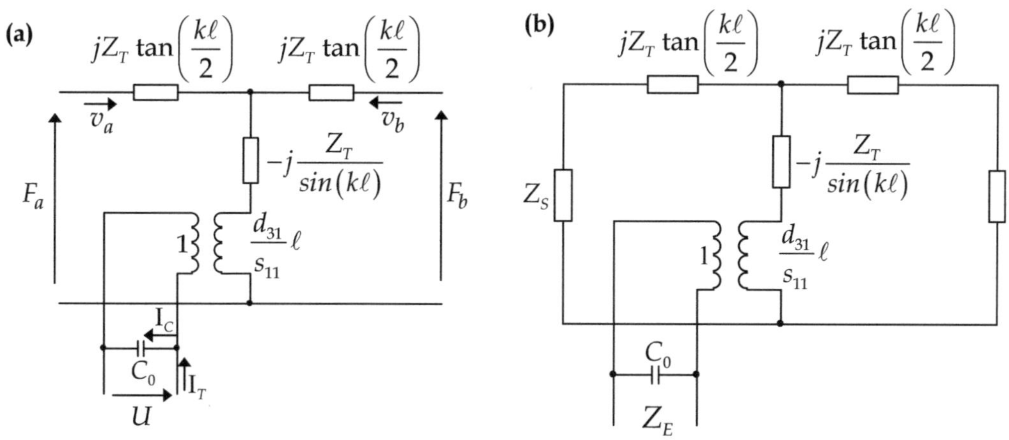

Fig. 4. (a) Piezoelectric transducer represented by an equivalent electromechanical circuit and (b) both acoustic ports loaded by the mechanical impedance of the host structure.

Analyzing the circuit in Figure 4 (b), we can obtain the equivalent electrical impedance between the terminals of the transducer, which is given by

$$Z_E = \frac{1}{j\omega C_0}\left\| jZ_T\left(\frac{s_{11}}{d_{31}\ell}\right)^2\left[\frac{1}{2}\tan\left(\frac{k\ell}{2}\right) - \frac{1}{sin(k\ell)} + \frac{Z_S}{j2Z_T}\right]\right. \tag{51}$$

According to Equation (51), there is a relationship between the mechanical impedance of the monitored structure and the electrical impedance of the piezoelectric transducer. Changes in the mechanical impedance of the structure due to damage result in a corresponding change in the electrical impedance of the transducer. Therefore, structural damage can be characterized by measuring the electrical impedance in an appropriate frequency range. This is the basic principle of damage detection discussed in the next section.

3.4 Damage detection

The comparison between the electrical impedance signatures of a PZT transducer unbonded and bonded to an aluminum beam in a frequency range of 10-40 kHz is shown in Figure 5. When the transducer is bonded to the structure, several peaks are observed in both the real part and the imaginary part signatures.

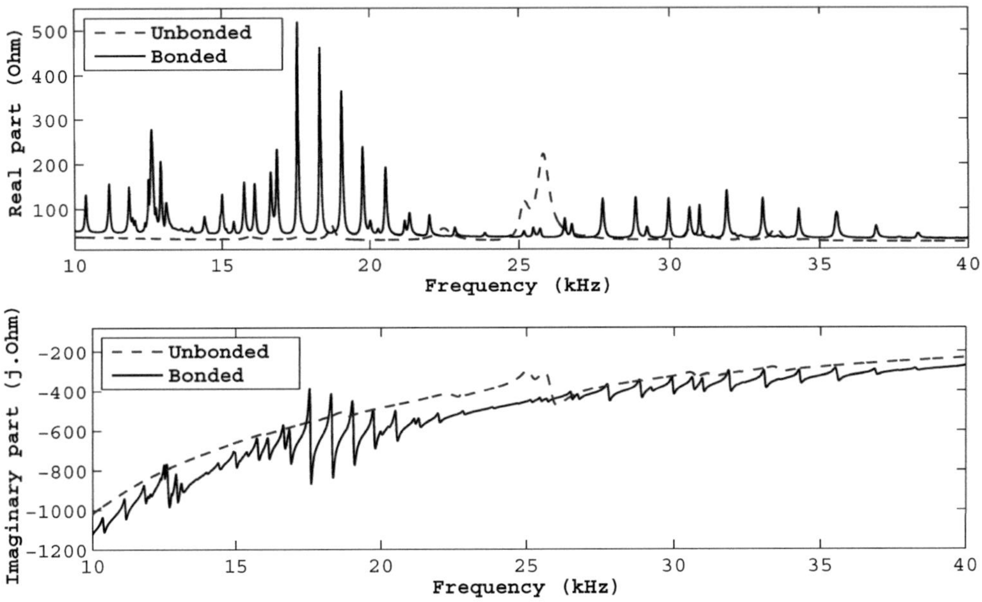

Fig. 5. Real part and imaginary part of the electrical impedance signatures in a frequency range of 10-40 kHz for a PZT transducer unbonded and bonded to an aluminum beam.

These peaks are related to the natural frequencies of the monitored structure. Changes in the natural frequencies either in frequency shifts or variations in the amplitude may indicate structural damage.

Usually, the characterization of damage is performed through metric indices by comparing two impedance signatures, where one of these is previously acquired when the structure is considered healthy and used as reference, commonly called the baseline. Thus, the electrical impedance is repetitively acquired and compared with the baseline signature.

Various indices have been proposed in the literature for damage detection, but the most widely used is the root mean square deviation (RMSD) which is based on Euclidian norm (Giurgiutiu & Rogers, 1998). Some changes in this index have been suggested by several researchers. One of the most used is given by

$$RMSD = \sum_{n}^{M} \sqrt{\frac{\left(Z_{n,d} - Z_{n,h}\right)^2}{Z_{n,h}^{\ 2}}} \tag{52}$$

where $Z_{n,h}$ and $Z_{n,d}$ are the electrical impedance (magnitude, real or imaginary part) for the host structure in healthy and damaged condition, respectively, measured at frequency n, and M is the total number of frequency components, which is related to the frequency resolution of the measurement system.

The index in Equation (52) should be calculated within an appropriate frequency range, which provides good sensitivity for damage detection. Generally, the suitable frequency range is selected experimentally by trial and error methods, but recently some researchers have proposed more efficient methodologies (Peairs et al., 2007; Baptista & Vieira Filho, 2010). In addition to selecting the appropriate frequency range, it is essential that the measurement system has a good sensitivity and repeatability to avoid either false negative or false positive diagnosis in detecting damage. The measurement systems based on virtual instrumentation are presented in the next section.

4. Electrical impedance measurement

Normally, the measurement of the electrical impedance, which is the basic stage of the EMI technique, is performed by commercial impedance analyzers such as the 4192A and 4294A from Hewlett Packard / Agilent, for example. Besides the high costs, these instruments are slow, making it difficult to use the technique in real-world applications, where it is required to use multiple sensors and to diagnose the structure in real-time. The conventional impedance analyzers use a pure sinusoidal wave at each frequency step, making a stepwise measurement under steady-state condition within an appropriate frequency range. Based on this principle, many researchers have developed alternative and low-cost systems for general impedance measurements. Usually, these systems are based on the volt-ampere method (Ramos et al., 2009) where the sinusoidal signal at each frequency step is supplied by a function generator or a direct digital synthesizer - DDS (Radil et al., 2008).

Steady-state measurement systems for specific applications in SHM have also been proposed. In the system proposed by Panigrahi et al. (2010), a function generator was used to excite gradually the piezoelectric transducer with pure sinusoidal signals at each frequency step and an oscilloscope was employed to measure the output response at each excitation frequency. This system is an improvement from a previous work developed by Peairs et al., (2004) where a fast Fourier transform (FFT) analyzer was used to obtain the electrical impedance in the frequency domain. Recently, Analog Devices developed a miniaturized high precision impedance converter, which includes a frequency generator, a DDS core, analog-to-digital converter (ADC) and digital-to-analog converter (DAC), a digital-signal-processor (DSP) integrated in a single chip (AD5933). This chip is used with a microcontroller and other required devices and can provide electrical impedance measurements with high accuracy. This chip has been used in SHM to develop compact and low-cost measurement systems. These new systems support wireless communication and several sensors through analog multiplexer, and can process data locally (Min et al., 2010; Park et al., 2009).

Although steady-state measurement systems provide results with high accuracy, the measurements usually take a long time because the frequency of the pure sinusoidal signal should be gradually increased step-by-step within the suitable range for damage detection. The time consumption may be very significant if a wide frequency range with many steps is required. Accordingly, in these new portable systems a wide frequency range with a narrow frequency step demands a large amount of data that can be difficult to be stored and processed locally or transmitted in a wireless mode. Moreover, these systems are difficult to assemble or require specific evaluation boards, which make them attractive mainly for specific SHM applications. For research and general applications, a simpler system should be developed.

We can design fast and easy to assemble measurement systems if the piezoelectric transducer is excited with a sweep signal and using virtual instrumentation. The concept of virtual instrumentation is shown in Figure 6.

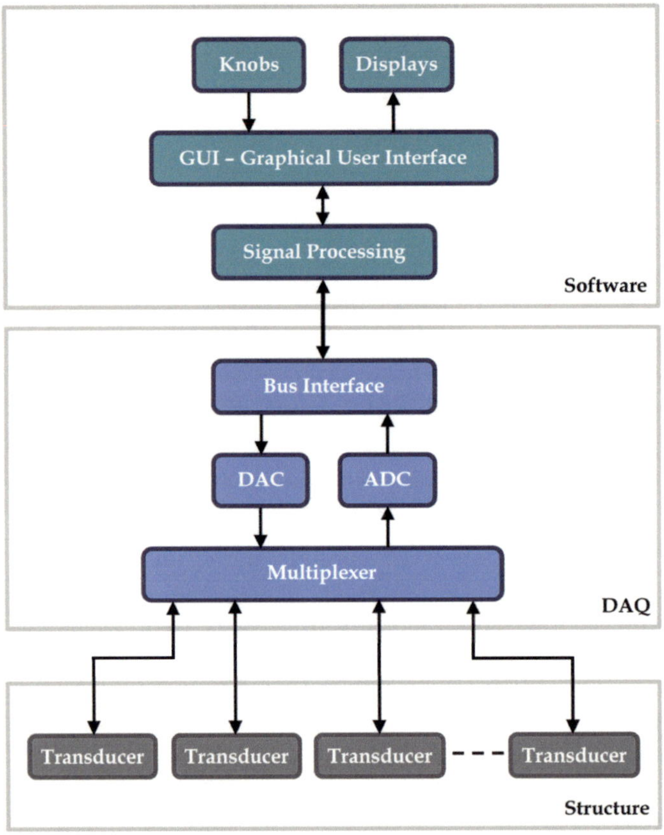

Fig. 6. Data acquisition and virtual instrumentation for structural health monitoring based on EMI methods.

The hardware consists mainly of ADC and DAC converters, multiplexer and bus interface, generally integrated into a single data acquisition (DAQ) device. The DAC generates the excitation signal in a frequency range selected by the user. The response signal of the transducer with information related to the structural health is acquired by the ADC, which must have a sampling rate of at least twice the maximum frequency in the excitation signal. The multiplexer allows the connection of multiple transducers and the bus interface provides the connection with a personal computer.

The control of the DAQ device and the signal processing is done through software. The graphical user interface (GUI) provides knobs and displays for adjusting the parameters of the signal generation and acquisition and the visualization of data, such as impedance signatures and metric indices. The virtual instrumentation makes the system very versatile. Adaptations, displays and knobs can be easily added if necessary.

The virtual instrumentation presented in this chapter is mainly based on LabVIEW (Laboratory Virtual Instrument Engineering Workbench), a graphical programming environment from National Instruments. However, other programming platforms can be used, such as Matlab from Mathworks that provides tools for data acquisition and is commonly found in research laboratories.

A basic LabVIEW program for signal acquisition and generation using a DAQ device is shown in Figure 7.

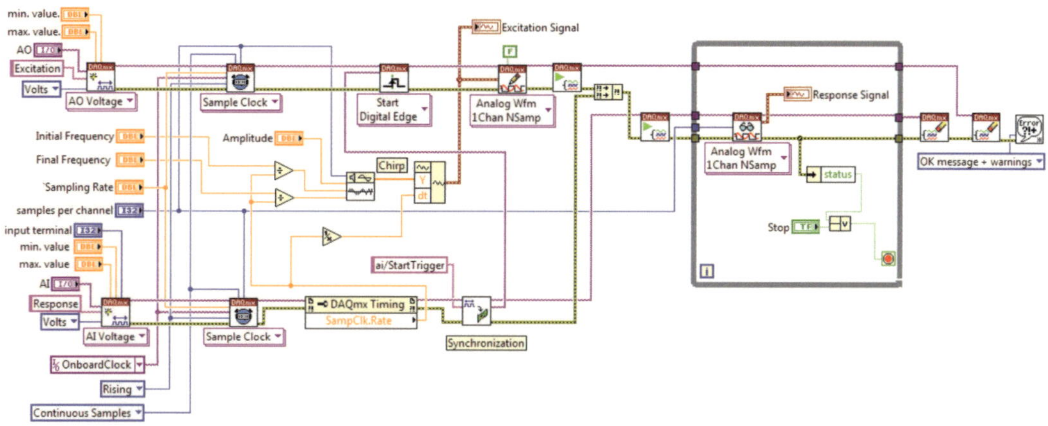

Fig. 7. Basic LabVIEW program for signal acquisition and generation.

The program sets through virtual channels the analog output (AO) and the analog input (AI) used for the excitation of the transducer and the acquisition of the corresponding response signal, respectively. The excitation is performed by a linear chirp signal, which makes a sweep from an initial frequency to a final frequency. The synchronization between the signal generation and acquisition is essential to ensure excellent repeatability between the measurements, which is an important feature in SHM to avoid incorrect diagnosis. The loop structure allows various cycles of excitation and response in applications where is required an average between the measurements for high accuracy.

The characterization of damage can be carried out by analyzing the electrical impedance signatures in the frequency domain or directly the response signal from the transducer in the time domain, as presented in the next sections.

4.1 Frequency domain analysis

The analysis in the frequency domain is the usual way to characterize damage and it is usually based on the FRF, as mentioned in the Section 2. The FRF is obtained from the excitation and response signals considered as input and output, respectively, of the system under test containing the structure and the transducer. Thus, from the FRF, we can calculate the electrical impedance in an appropriate frequency range.

The measurement system proposed by Baptista & Vieira Filho (2009) is based on the concept shown in Figure 6 and uses a low-cost DAQ device with maximum sampling rate of 250 kS/s, limiting the impedance measurement up to 125 kHz, limiting the impedance measurement up to frequency of 125 kHz, although other devices with higher sampling rate can be used without significant changes in the software. The software was developed in LabVIEW with the basic configuration shown in Figure 7. Besides the software, the hardware is very simple and uses only a common resistor in addition to DAQ device to connect the transducer. The system diagram is shown in Figure 8.

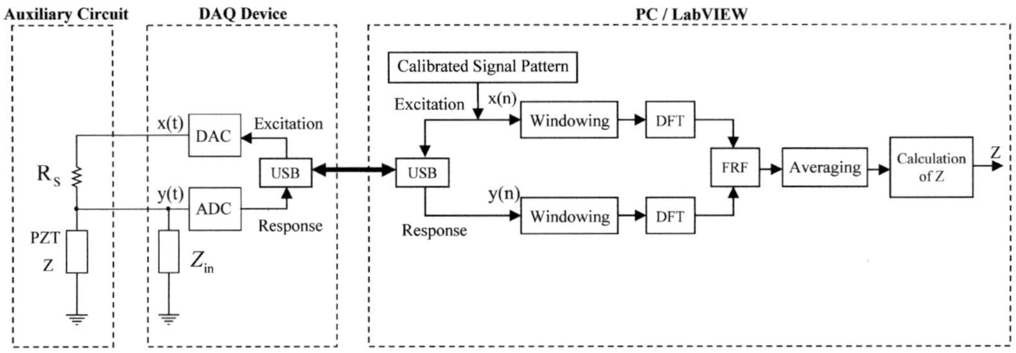

Fig. 8. Example of measurement system based on DAQ device and virtual instrumentation (Baptista & Vieira Filho, 2009).

In Figure 8, $x(t)$ and $y(t)$ represent the analog signals of excitation and response, respectively, and $x(n)$ and $y(n)$ represent the digital signals. The measurements are based on averaged FRF ($\overline{H}$) obtained from discrete Fourier transform (DFT) of the excitation and response signals. Thus, the electrical impedance (Z_E) in the frequency domain is given by

$$Z_E[n] = \frac{\overline{H[n]} \cdot R_S\left(r + Z_{in}[n]\right)}{Z_{in}[\mathrm{n}] - \overline{H[n]}\left(R_S + r + Z_{in}[\mathrm{n}]\right)} \tag{53}$$

where the resistor R_S is a current limiter, r is the resistance of the cable used to connect the DAQ device and the PZT sensor, and Z_{in} is the input impedance of the DAQ device and n is de frequency of the chirp signal ranging from an initial low value to a final high value as defined in the basic program shown in Figure 7.

This system was tested in a PZT transducer bonded to an aluminum beam and the results were compared with the measurements obtained using a conventional impedance analyzer 4192A from Hewlett Packard. The comparison between the two electrical impedance signatures in a frequency range of 30-50 kHz is shown in Figure 9. The similarity between the two signatures indicates the accuracy and feasibility of the system based on virtual instrumentation. The discrepancy between the two measures is less than 4%. Besides the good accuracy, the measurement system provides fast measurements, versatility and low-cost compared to the conventional impedance analyzers.

Improvements such as multiple sensors and real-time diagnosis can be easily included. The system diagram shown in Figure 10 has these features (Baptista et al., 2011). As in the previous system, the excitation signal *x(t)* is generated by the DAC which has a output impedance R_{out} and the corresponding response signal *y(t)* from each PZT sensor is acquired by the ADC which has a input impedance Z_{IN} constituted by a high resistance connected in parallel with a capacitance. The low cost DAQ devices have only one ADC and each analog input is routed through an onboard multiplexer (MUX). The samples *n* of the excitation and the response signals in the discrete form ($x[n]$ and $y[n]$) are synchronized by software through the onboard clock.

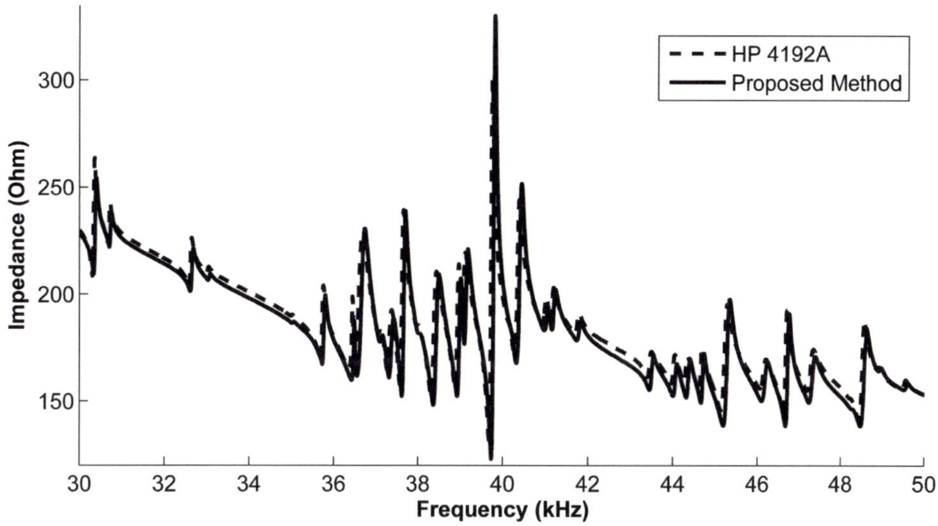

Fig. 9. Comparison between the electrical impedance signatures obtained with a conventional impedance analyzer and the system based on virtual instrumentation.

The resistors R_L are current limiters and are connected in series with the sensors. The sensors have common ground, which facilitates the installation of multiple sensors in metallic structures. This common ground should be connected to the DAQ ground.

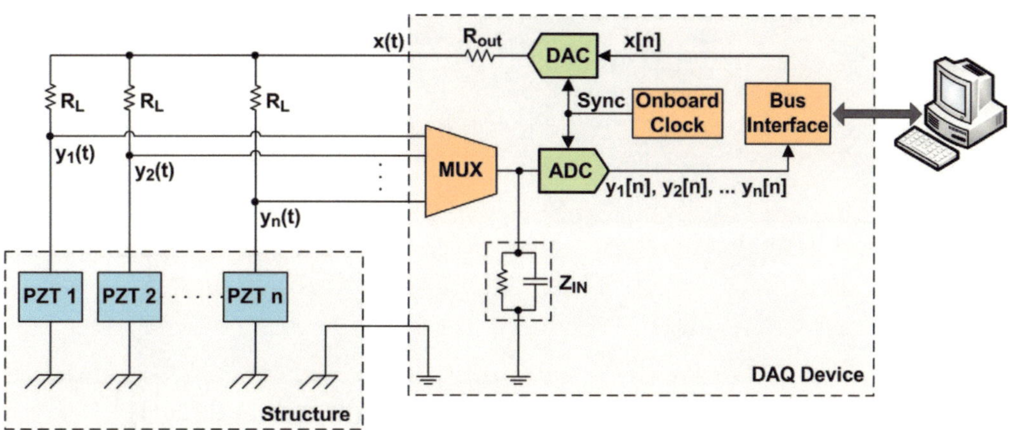

Fig. 10. Improved measurement system including multiple sensors and real-time diagnosis (Baptista et al., 2011).

Typically, the output impedance R_{out} is very low on the order of tenths of Ohm. If the resistances R_L are much higher than the output impedance, i.e., $R_L >> R_{out}$, the excitation signal $x(t)$ in the analog output of the DAQ device can be considered constant in relation to the variations in the electrical impedance of the sensors. Consequently, any mutual interference between the acquisition channels can be neglected and the variations in the response signals $y_1(t)$, $y_2(t)$, …, $y_n(t)$ from each sensor are caused only by the mechanical properties of the host structure or other environmental conditions. The piezoelectric transducers, especially those made of thin PZT ceramics, require low voltage and low current for the excitation signal, so that the piezoelectric effect is linear (Sun et al., 1995; Baptista et al., 2010). Thus, the resistance R_L may be on the order of some thousands of Ohm and the condition $R_L >> R_{out}$ is easily satisfied.

As in the previous system, the software was implemented in LabVIEW. Figure 11 shows a user-friendly interface that allows the adjustment of the parameters of the signal generation and signal acquisition. A display shows the baseline and the current impedance signatures of the selected sensor and vertical bars indicate the level of damage. In addition, a panel displays a 3D model of the host structure, a LabVIEW tool known as *sensor mapping*. In this 3D model, virtual sensors are placed in the same positions that they have in the real-world structure, changing the color of the model according to the intensity and in the regions indicated by the metric indices. This feature gives a reasonable suggestion of the damage location, especially in large structures with many sensors. Note instead of the RMDD index, it was used the CCDM index, which is based on the correlation coefficient.

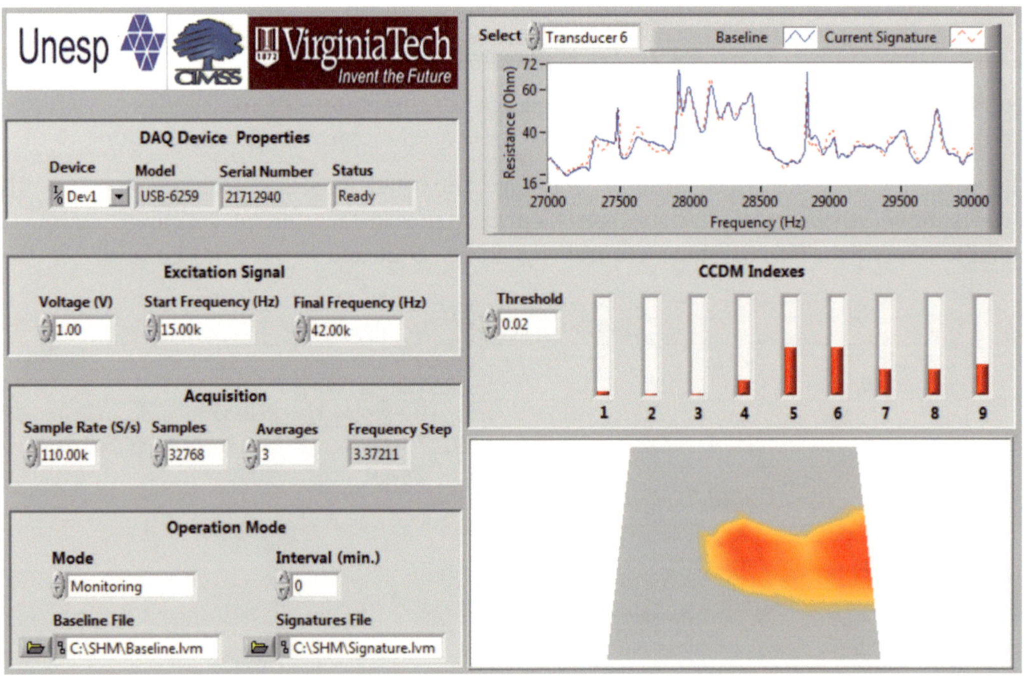

Fig. 11. User-friendly interface of the measurement system (Baptista et al., 2011).

The system was evaluated in an aluminum plate with nine transducers and the DAQ device used was the model USB-6259 from National Instruments. This model has 16 differential analog inputs and a maximum sampling rate of 1.00 MS/s in the aggregated mode. Thus, for the acquisition of the signals from the nine sensors the sampling rate was set to 110 kS/s with 32768 samples. The results show conclusively that the system has good repeatability, sensitivity to detect damage and provides fast measurements. For this configuration, the measurement of the nine sensors and the presentation of the results are completed in less than 2 seconds in a PC laptop with medium performance.

The systems presented previously are based on the FRF, where structural damage is detected by analyzing the electrical impedance signatures in the frequency domain. However, the characterization of damage can be performed in a simpler method directly in the time domain, as discussed in the next section.

4.2 Time domain analysis

In this section, we show that the time response of a piezoelectric transducer provides information on the electromechanical impedance variation when a monitored structure is damaged (Vieira Filho et al., 2011). The time domain approach changes the paradigm of SHM systems based on EMI and the results are similar to those obtained using electrical impedance measurements in the frequency domain. The efficiency of the time domain approach was demonstrated through experiments using an aluminum plate. The results using both the FRF and the time response were obtained and compared.

In the time domain approach, the analysis of only the time response *y(t)* of the transducer using the system presented in Figure 8 is sufficient to detect damage. From Figure 8, the time response of the transducer *y(t)* in relation to the excitation signal *x(t)* can be obtained through an inverse Fourier or Laplace transform according to the following equation

$$Y = \frac{Z_E}{Z_E + R_S} X \tag{54}$$

where Z_E is the electrical impedance of the transducer.

From a practical point of view, the inverse transform is not necessary because the time response is directly obtained. However, the response signal *y(t)* changes according to the electrical impedance Z_E and the input signal *x(t)*. Considering an input signal with constant amplitude and frequency, the response signal *y(t)* will change only if the electrical impedance Z_E changes, which according to Equation (51) occurs when the structure suffers any type of damage and its mechanical impedance Z_S changes . In this case, the signal *y(t)* could be directly related to the health condition of the monitored structure. If *x(t)* is a pure sine wave signal with peak amplitude V_{px} and fixed frequency ω_x, it can be shown (Radil et al., 2008) that the approximate magnitude of the electrical impedance Z_E is given by

$$|Z_E| \cong \frac{V_{py}}{V_{px} - V_{py}} R_S \tag{55}$$

where V_{py} represents the amplitude peak of the response signal *y(t)*.

For a complete characterization of Z_E, it is possible to compute both the real and the imaginary parts. This is quite direct if the phase difference between *x(t)* and *y(t)* is known.

However, the magnitude of the impedance might assure good sensitivity as well and this will be shown in the example results. So, considering a constant peak value of $x(t)$, the response $y(t)$ will be modified according to any variation of Z_E, which affirms that $y(t)$ is function of Z_E. This approach is enough to detect damage because it is sensitive to any structural change. Furthermore, an efficient SHM method based on the EMI does not have to measure the electrical impedance itself, but just measure its variation. This new methodology is called here the time electrical impedance (TEI).

An experiment using an aluminum plate of 500 x 300 x 2 mm with four PZT transducers was carried out to validate the TEI method (Vieira Filho et al., 2011). Figure 12 presents the time response signal $y(t)$ in both healthy and damage conditions.

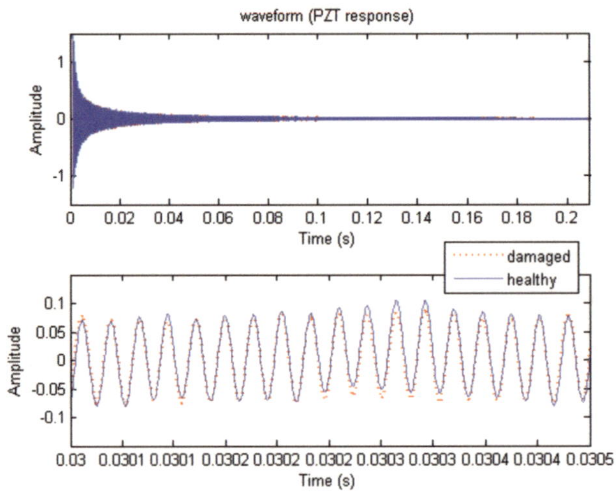

Fig. 12. Time response signal $y(t)$ in health and damage conditions.

We can observe that both responses are close. However, a significant difference can be observed if a subtraction between time responses ($y_b[n]-y_m[n]$) is carried out, where y_b is the baseline, y_m is the updated time response and n is the sample. This operation was carried out first between the baseline and the time response in the healthy condition and then between the baseline and the time response in the damaged condition. The results are presented in Figure 13 and they show that this operation gives suitable information on the structure's condition. As a result, it is expected that these differences could be detected using metric indices, such as RMSD.

The TEI method was implemented and the results were compared to the ones obtained using the traditional EMI based on the FRF. Thus, the RMSD values were obtained for both TEI and FRF. It is important to observe that although the absolute values of the indices are not the best method for comparing the results, they are interesting for the purpose of evaluating each method separately. However, since the goal is to compare TEI and FRF, the indices are presented using the ones obtained in healthy condition as reference (called normalized here). Figure 14 shows the normalized RMSD values for (a) TEI and (b) for FRF obtained using a plate with four PZT patches and damage simulated at three different positions. According to the experimental results, the RMSD values obtained in the time domain are significantly higher than those obtained from FRF. In the time domain, the variation in these values for the damaged structure in relation to the healthy condition reaches a factor of 45 times greater. On the other hand, in the frequency domain this variation is only about 12 times greater.

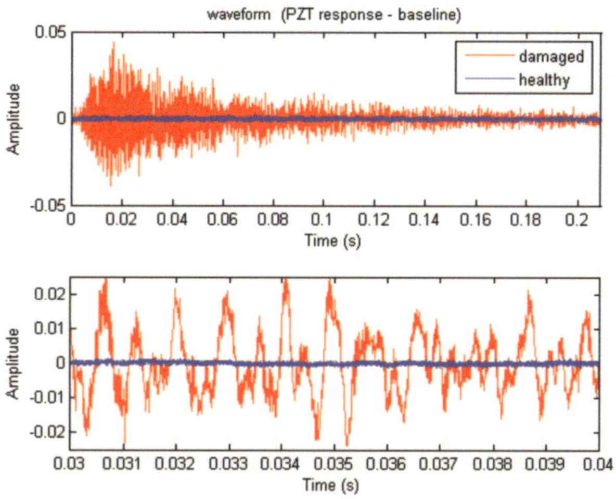

Fig. 13. Difference between the baseline and updated time response signals.

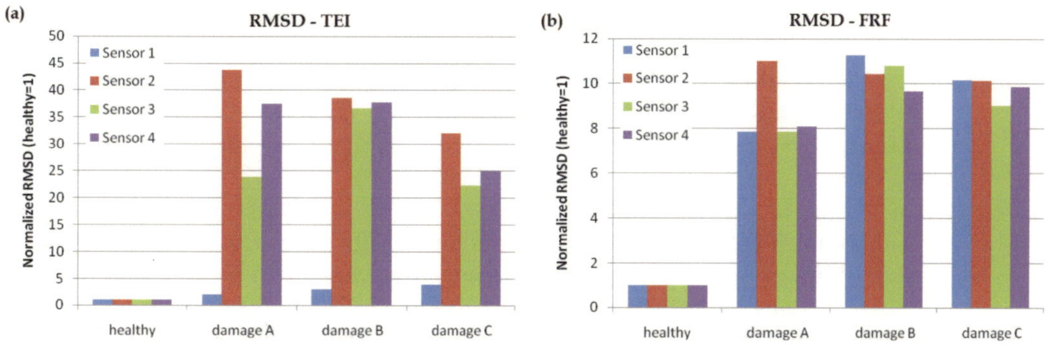

Fig. 14. Normalized RMSD values for (a) TEI and (b) for conventional FRF method.

Therefore, the results indicate conclusively that the characterization of damage in the time domain is feasible and has excellent sensitivity. Since the analysis is carried out directly from the time response signal and it is not required to compute the DFT, the TEI is simpler than the conventional EMI method.

5. Conclusion

In this chapter, we have presented the basic principle of the electromechanical impedance technique for detecting damage in structural health monitoring. The measurement of the electrical impedance of piezoelectric transducers, which is the basic stage of the technique, was addressed focusing the virtual instrumentation. The analysis in the frequency domain and the time domain were presented. The experimental results show conclusively that the measurement systems based on virtual instrumentation are feasible and efficient for both methods of analysis.

It is important to note that the equivalent electromechanical circuit and the experimental results presented here were obtained for the PZT ceramics. However, these ceramics are brittle and in some applications it is advantageous to use MFC transducers, which are more flexible. In addition, the electromechanical impedance signatures are significantly sensitive to temperature variations. Therefore, the measurements systems should include some compensation method to correct the changes in the impedance signatures for practical applications under temperature variations.

6. Acknowledgment

The authors would like to thank the Center for Intelligent Material Systems and Structures, Virginia Tech, and INCT-EIE. This work was partially supported by the Capes Foundation, Ministry of Education of Brazil, grant numbers BEX 0125/10-5 and BEX 3634/09-4.

7. References

Baptista, F.G. & Vieira Filho, J. (2009). A New Impedance Measurement System for PZT Based Structural Health Monitoring. *IEEE Transactions on Instrumentation and Measurement*, Vol. 58, No. 10, (October 2009), pp. 3602-3608, ISSN 0018-9456

Baptista, F.G. & Vieira Filho, J. (2010). Optimal Frequency Range Selection for PZT Transducers in Impedance-Based SHM Systems. *IEEE Sensors Journal*, Vol. 10, No. 8, (August 2010), pp. 1297-1303, ISSN 1530-437X

Baptista, F.G.; Vieira Filho, J. & Inman, D.J. (2010). Influence of Excitation Signal on Impedance-Based Structural Health Monitoring. *Journal of Intelligent Material Systems and Structures,* Vol. 21, No. 14 (November 2010), pp. 1409-1416, ISSN 1045-389X

Baptista, F.G.; Vieira Filho, J. & Inman, D.J. (2011). Real-Time Multi-Sensors Measurement System With Temperature Effects Compensation for Impedance-Based Structural Health Monitoring. *Structural Health Monitoring,* DOI: 10.1177/1475921711414234 (published online before print), pp. 1-14, ISSN 1475-9217

Cawley, P. (1997). Long Range Inspection of Structures Using Low Frequency Ultrasound, *Proceedings of Structural Damage Assessment Using Advanced Signal Processing Procedures*, pp. 1-17, University of Sheffield, Sheffield

Giurgiutiu, V. & Rogers, C.A. (1998). Recent Advancements in the Electro-Mechanical (E/M) Impedance Method for Structural Health Monitoring and NDE, *Proceedings of 5th Annual International Symposium on Smart Structures and Materials*, Vol.3329, pp. 536-547, SPIE, San Diego, USA

Gyekenyesi, A.L.; Martin, R.E.; Sawicki, J.T. & Baaklini, G.Y. (2005). Damage Assessment of Aerospace Structural Components by Impedance Based Health Monitoring. *NASA Technical Memorandum,* TM—2005-213579, Hanover, Available from http://gltrs.grc.nasa.gov

Kessler, S.S.; Spearing, S.M.; Atala, M.J.; Cesnik, C.E.S. & Soutis, C. (2002). Damage Detection in Composite Materials Using Frequency Response Methods. *Composites Part B: Engineering*, Vol. 33, No. 1, (January 2002), pp. 87-95, ISSN 1359-8368

Kossoff, G. (1966). The Effects of Backing and Matching on The Performance of Piezoelectric Ceramic Transducers. *IEEE Transactions on Sonics and Ultrasonics*, Vol. 13, No. 1, (March 1966), pp. 20-30, ISSN 0018-9537

Meitzler, A.H. et al. (1987). IEEE Standard on Piezoelectricity: An American National Standard. Std 176, 66 p., IEEE-ANSI, New York, USA

Min, J.; Park, S.; Yun, C.B. & Song, B. (2010). Development of Multi-Functional Wireless Impedance Sensor Nodes for Structural Health Monitoring, *Proceedings of SPIE Sensors and Smart Structures Technologies for Civil, Mechanical, and Aerospace Systems 2010*, vol. 7647, pp. 764728-1–764728-8, San Diego, CA, USA

Panigrahi, R.; Bhalla, S. & Gupta, A. (2010) A Low-Cost Variant of Electro-Mechanical Impedance (EMI) Technique for Structural Health Monitoring. *Experimental Techniques*, Vol. 34, No. 2, (March 2010), pp. 25–29, ISSN 1747-1567

Park, S.; Shin, H.H. & Yun, C.B. (2009) Wireless Impedance Sensor Nodes for Functions of Structural Damage Identification and Sensor Self-Diagnosis. *Smart Materials and Structures*, Vol. 18, No. 5, (May 2009), pp. 055001, ISSN 0964-1726

Peairs, D.M.; Park, G. & Inman, D.J. (2004). Improving Accessibility of the Impedance-Based Structural Health Monitoring Method. *Journal of Intelligent Material Systems and Structures*, Vol. 15, No .2, (February 2004), pp. 129–139, ISSN 1045-389X

Peairs, D.M.; Tarazaga, P.A. & Inman, D.J. (2007). Frequency Range Selection for Impedance-Based Structural Health Monitoring. *Journal of Vibration and Acoustics*, Vol. 129, No. 6, (December 2007) pp. 701-719, ISSN 1048-9002

Radil, T.; Ramos, P.M. & Serra, A.C. (2008). Impedance Measurement with Sine-Fitting Algorithms Implemented in a DSP Portable Device. *IEEE Transactions on Instrumentation and Measurement*, Vol. 57, No. 1, (January 2008), pp. 197-204, ISSN 0018-9456

Ramos, P.M.; Janeiro, F.M.; Tlemçani, M. & Serra, A.C. (2009). Recent Developments on Impedance Measurements with DSP-Based Ellipse-Fitting Algorithms. *IEEE Transactions on Instrumentation and Measurement*, Vol. 58 No. 5, (May 2009) pp. 1680-1689, ISSN 0018-9456

Royer, D. & Dieulesaint, E. (2000). Elastic Waves in Solids II: Generation, Acousto-Optic Interaction, Applications, Vol. 2, 446 p., Springer, Berlin

Rytter, A. (1993). *Vibration Based Inspection of Civil Engineering Structures*. Department of Building Technology and Structural Engineering, Aalborg University, 193 p., Denmark

Sohn, H.; Farrar, C.R.; Hemez, F.M.; Shunk, D.D.; Stinemates, D.W.; Nadler, B.R. & Czarnecki, J.J. (2004). A review of Structural Health Monitoring Literature: 1996–2001. *Los Alamos National Laboratory Report*, LA-13976-MS, Available from http://www.lanl.gov

Sun, F.; Chaudhry, Z.; Liang, C. & Rogers, C.A. (1995). Truss Structure Integrity Identification Using PZT Sensor-Actuator. *Journal of Intelligent Material Systems and Structures*, Vol. 6, No. 1, (January 1995), pp. 134–139, ISSN 1045-389X

Vieira Filho, J.; Baptista, F.G.; Farmer, J. & Inman, D.J. (2011). Time-Domain Electromechanical Impedance for Structural Health Monitoring, *Proceedings of 8th International Conference on Structural Dynamics*, Leuven, Belgium, 4-6 July 2011

Wang, M.L.; Satpathi, D. & Heo, G. (1997). Damage Detection of a Model Bridge Using Modal Testing, *Proceedings of International Workshop on Structural Health Monitoring*, pp. 589-600, DEStech Publications, Stanford, California, USA